Geometry Juniors
by Ed Southall

Illustrated by Nicole Lane

© The Mathematical Association 2020
All rights reserved.

First published in 2020 by
The Mathematical Association
ISBN 978-1-911616-05-4

Printed and bound in Great Britain
by Blissetts Bookbinders

Copyright Statement
Photocopying of the material in this book for use in a single
educational institution that purchased the book is allowed as long
as it is used for purely educational purposes that are not
commercial in nature.

Contents

Introduction 1

Chapter 1 – Four thoughts 9

Chapter 2 – Properties of Shapes 31

Chapter 3 – Symmetry 53

Chapter 4 – Cubes 75

Chapter 5 – Coordinates 87

Chapter 6 – Nets 109

Chapter 7 – Angles 131

Chapter 8 – Areas 143

Chapter Notes 154

Introduction

This book is intended as a collaborative read between a parent and child, or teacher and students. It is deliberately designed to promote mathematical conversations about geometry. Often the intent of a diagram is merely to draw out what children understand and how they interpret shapes, properties and the limits of their knowledge. Whilst there are questions and prompts, these are not necessarily where your conversation will go, but are intended as a way to get children to think deeper about what they know and what they can find out. Some chapters are more prescriptive than others - the angles chapter near the end of the book for example, has very specific tasks to achieve, whereas several earlier chapters are more open and goal-free.

To demonstrate how conversations might develop with a child or class, I have provided an example of a conversation with a nine-year-old (J) on the following pages.

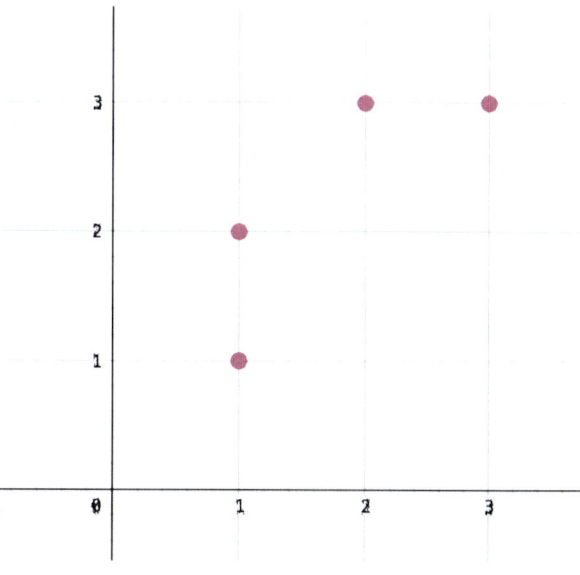

Me: Have a look at the points on this coordinate grid. Using the points, can you draw a trapezium?

J: Yes, I can see it.

(J draws the shape below with his finger)

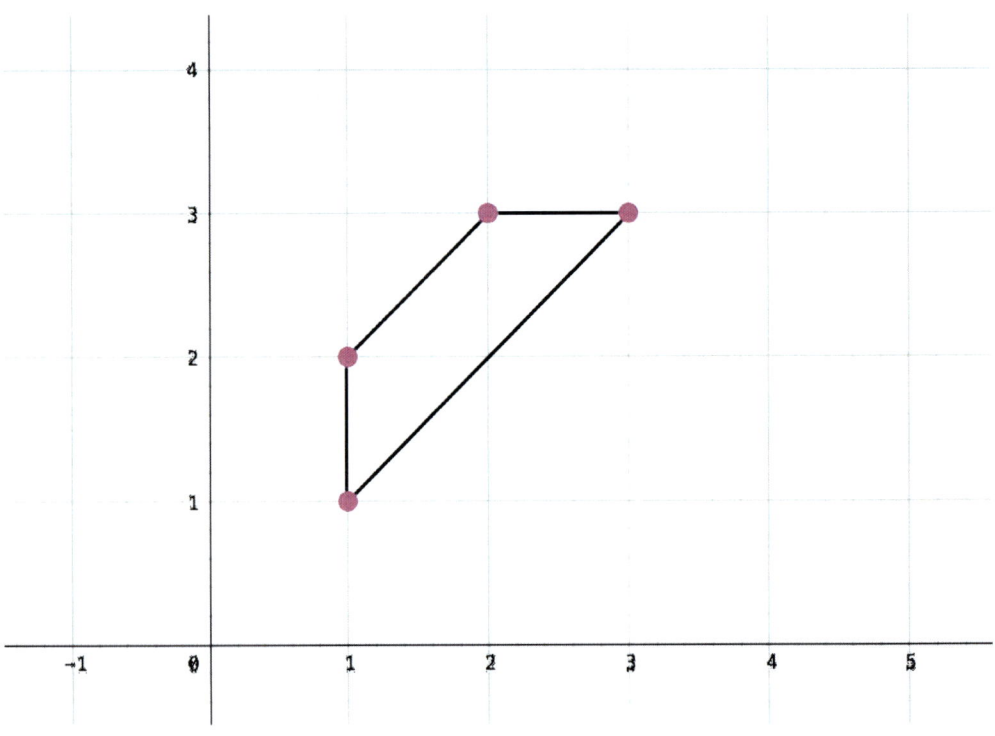

Me: Brilliant! Well spotted. How do you know that this is a trapezium?

J: Because it looks like a trapeze. There are two ropes and two bars.

Me: Is there anything special about the 'bars'?

J: They never meet, so they're parallel.

Me: Can you make a right-angle triangle by adding in an extra point?

J: Yes:

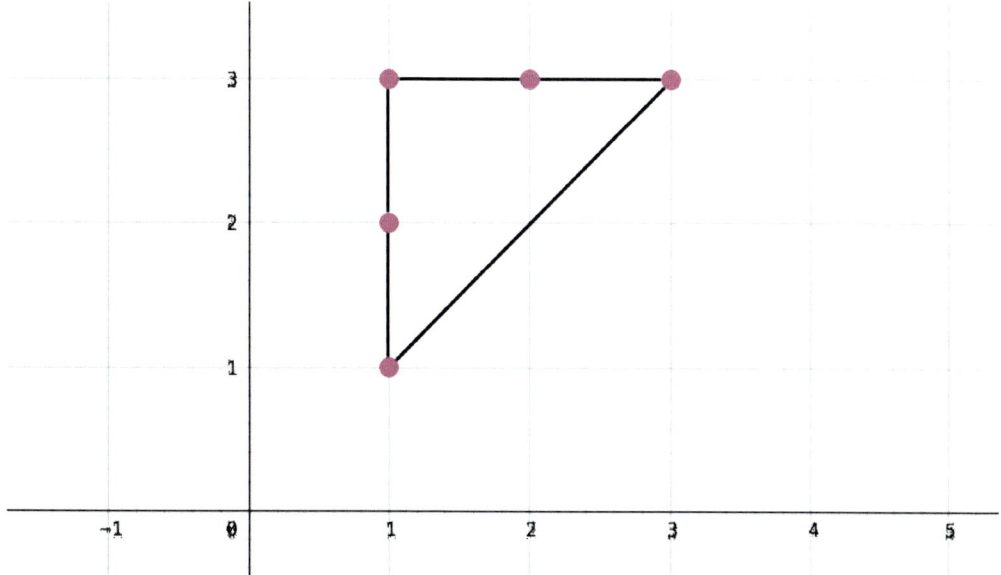

J: But it's isosceles as well as a right-angled triangle.

Me: Ok, last one: can you add a point to make a parallelogram?

J: Yes:

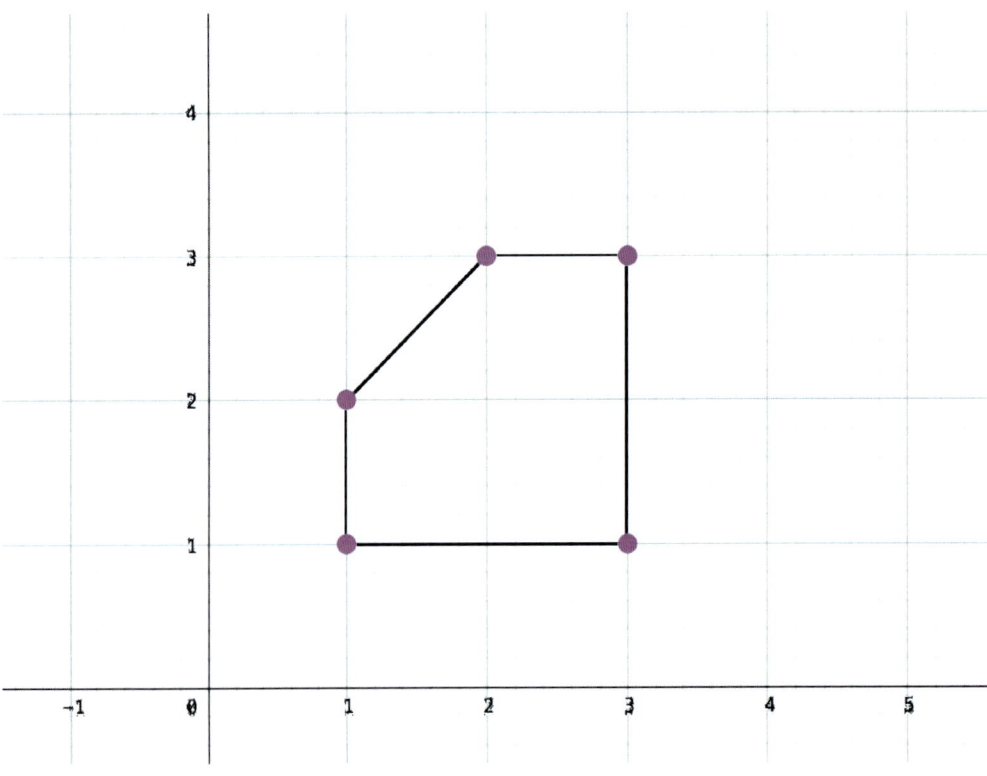

Me: Interesting, what makes this a parallelogram?

J: A parallelogram has an even number of sides, with pairs that are parallel.

Me: Let's count the sides.

J: Oh, this has five sides, it's not a parallelogram. I'll try again:

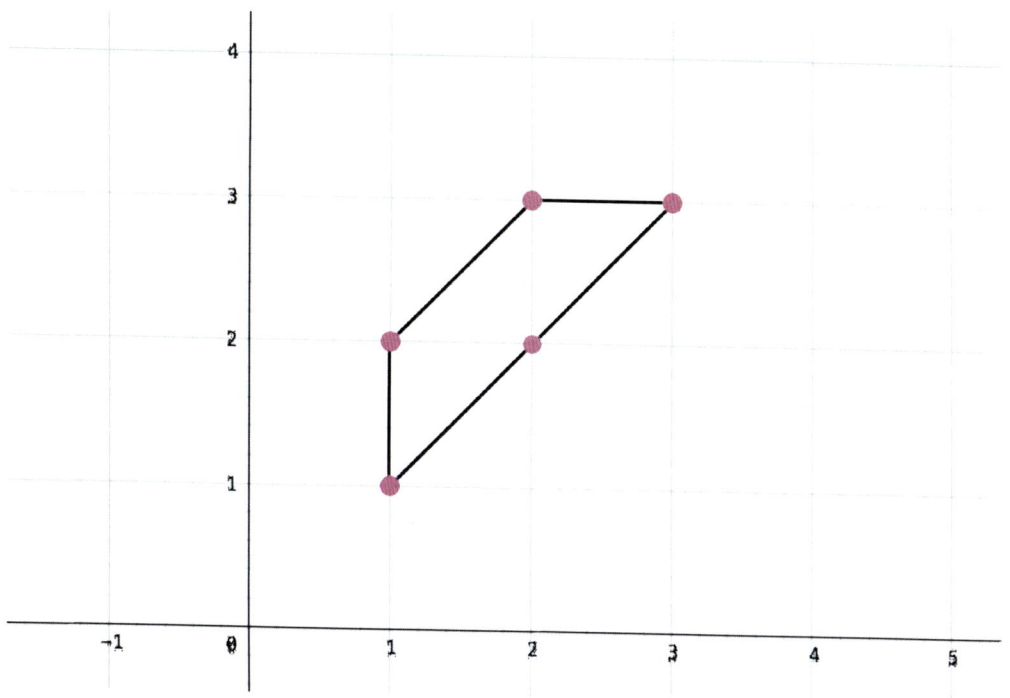

J: This is a parallelogram because it has an even number of sides and they are parallel.

Me: Are there two pairs of parallel sides?

J: No, just one pair... so then it's not a parallelogram. It can't be done then, you need two more points, not one.

Me: What if I put in a point here:

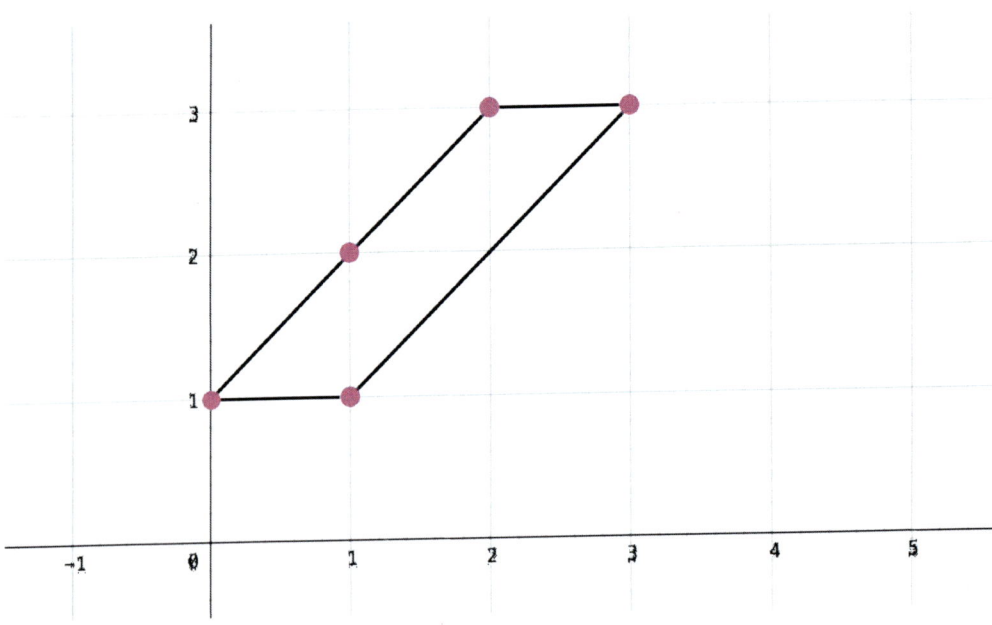

J: Oh, so that's a parallelogram. It has an even number of sides and pairs of sides are parallel.

Me: Well done! Can you have a parallelogram with six sides?

J: Yes, because it has even sides.

Me: I'm afraid not! Parallelograms are a kind of quadrilateral; they have to have exactly four sides!

The conversation enabled me to identify a couple of misconceptions about parallelograms and allowed the student to think deeply about what defines different types of shapes whilst problem solving. I purposefully chose not to correct the student initially, but instead enquired about their reasoning first to help me understand exactly what it was they believed to be true.

I hope that you find this book useful and enjoyable either at home or in the classroom.

Chapter 1
Four thoughts

Thinking Corner

> What do these shapes have in common?
>
> What else?
>
> Can you put your finger across to cut each shape in half?

Fact:
Oval means 'egg shaped'.

Thinking Corner

> What is different about each shape?
>
> How many squares can you see?
>
> What makes a rectangle a rectangle?

Fact:

Squares can be called regular quadrilaterals because all of their sides are equal in length and all of their angles are equal too.

Thinking Corner

Can you name each shape?

Which shape looks like it has the largest area?

Are any of these shapes regular?

Fact:

The simplest names for most shapes are based on the number of sides that they have, not by the way they look.

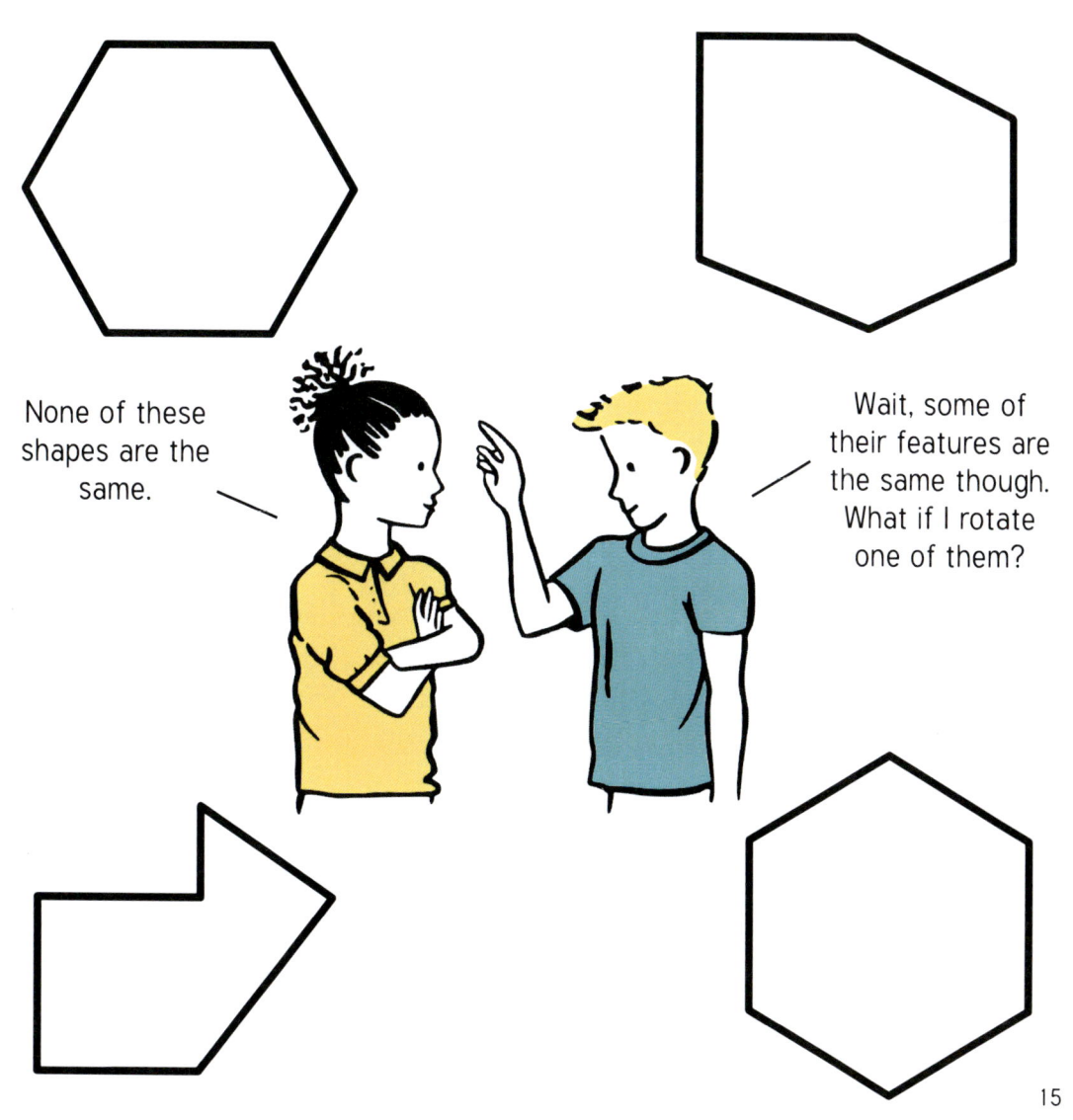

Thinking Corner

> Do any of these shapes have anything in common?
>
> Can you describe their angles?
>
> Can you categorise these shapes in different ways?

Fact:

The more sides a regular shape has, the greater their internal angles.

Thinking Corner

> Which shapes could be parts of circles?
>
> How could you halve them?
>
> Which one has the smallest perimeter?

Fact:

Different slices of a circle have different names: Semi-circles, sectors and segments!

Thinking Corner

Can you see any parallel lines?

How is each shape different?

Can you imagine each shape as two triangles stuck together?

How?

Fact:
The word 'trapezium' comes from the Greek for 'table'.

Thinking Corner

> How many of these shapes include an acute angle?
>
> Pick two shapes — how are they similar?
>
> What are their differences?

Fact:

Many shape names end in 'gon' meaning 'angled' and start with a number in Greek. For example, decagon means 'ten angles'.

Thinking Corner

What are the names for each of these triangles?

Which shape has the smallest angle?

What might we add to a diagram to show a triangle has equal sides?

Fact:

'Tri' in triangle means 'three'. Can you think of other words that use 'tri' in them meaning 'three'?

Thinking Corner

Are these triangles all the same?

Do they have the same perimeters?

Do they have the same angles?

Fact:

Two equilateral triangles can form a parallelogram if you put them together, and six equilateral triangles can form a regular hexagon!

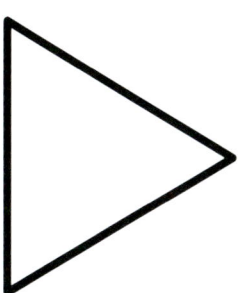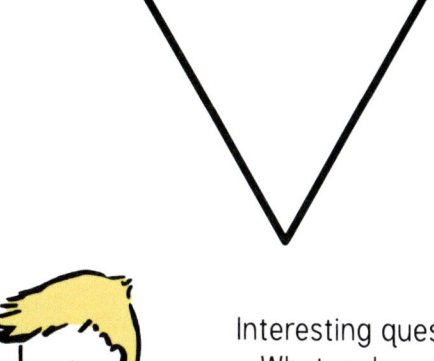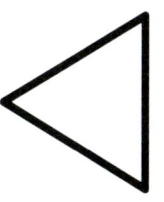

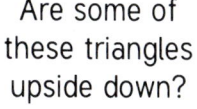 Are some of these triangles upside down?

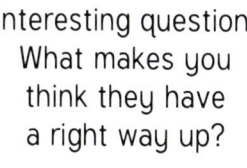 Interesting question! What makes you think they have a right way up?

Thinking Corner

Can you fit some of these shapes together to make new shapes?

Can you design two other shapes that fit together to make a new shape?

Fact:

The larger shapes in this picture are called a 'dart' and an 'annulus'.

Chapter 2
Properties of Shapes

Thinking Corner

What are the properties of each of these shapes?

How can you tell which one is a circle?

Can we use the same name to describe any of them?

Fact:
One of these shapes is called a discorectangle!

Thinking Corner

What properties do they all have in common?

Which of these shapes is isosceles?

Can we use other names to describe some of these types of triangles?

Fact:

Matching dashes on the sides of a shape means that they are the same length as each other.

Thinking Corner

If two triangles have an identical angle, are they the same?

Can two triangles of different sizes have the same angles?

What shapes can I make if I fit two right angle triangles together?

Fact:

The 'right' in 'right angle triangle' means 'upright', referring to the vertical perpendicular to a horizontal base.

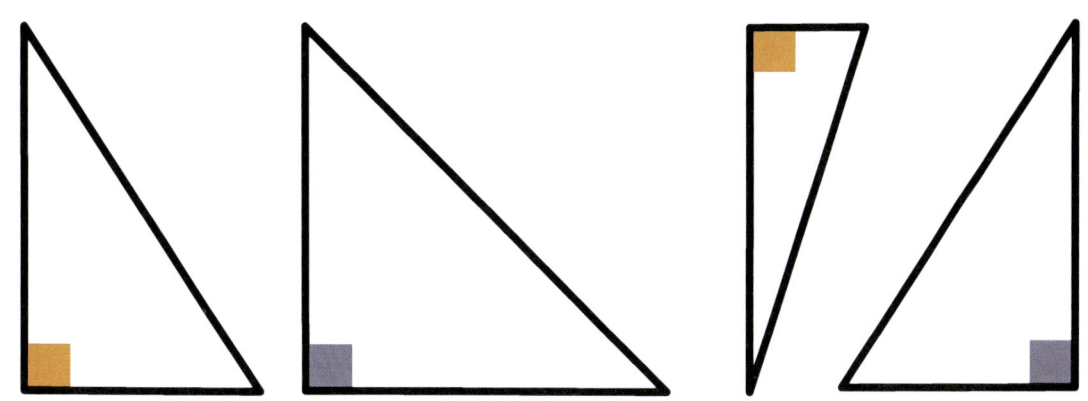

Thinking Corner

> What common properties do these shapes have?
>
> What properties are different?
>
> What do you notice about the angles inside each shape?

Fact:

We usually call a shape by its most specific name, but more general names are okay too. For example, a rectangle is a parallelogram.

Thinking Corner

> Which of these look like a parallelogram?
>
> What are the properties of a parallelogram?
>
> Is there anything special about their angles?

Fact:
We usually highlight parallel sides by drawing arrowheads on them.

Thinking Corner

What names could you give to these shapes?

What is the most specific name?

What is different about each shape?

Fact:

The orientation of a shape can make identification confusing. Their properties don't change if they're rotated though.

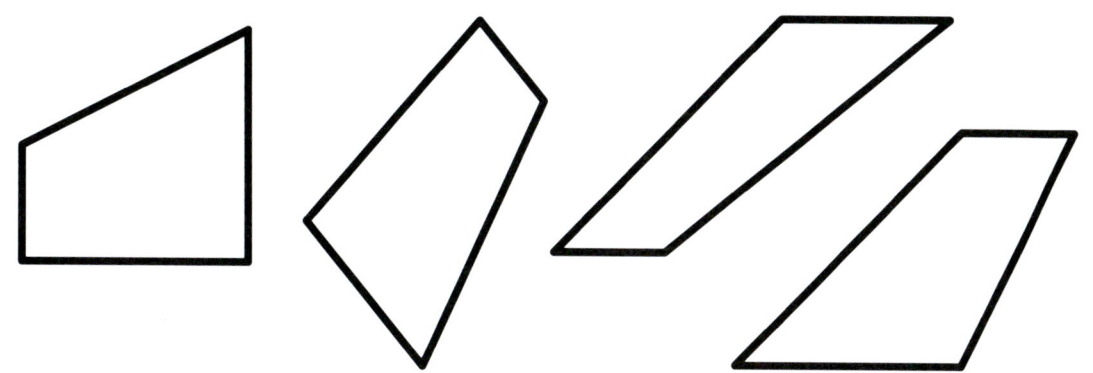

Thinking Corner

How can you show that these are all hexagons?

Do any of these shapes tessellate?

Can you figure out the size of any angles in these shapes?

Fact:

If two shapes have the same number of sides they will have a number of shared properties whether they are regular or irregular.

Thinking Corner

What are the names of each of these shapes?

What shared properties do they have?

Can some have more than one name?

Fact:

It's tempting to call a rhombus a 'diamond' but that's not its mathematical name.

Thinking Corner

> Do these shapes share any properties?
>
> What's tricky about the shape with a hole in it?
>
> What do you notice about the angles for these shapes?

Fact:
You can spot lots of shapes all around you.
For example, a 'STOP' road sign is a regular octagon.

Thinking Corner

What are the names of each of these overlapping regular shapes?

Each shape has an extra angle in it. How many more degrees are added each time?

How many straight lines can you see here?

Fact:

For regular shapes with an even number of sides, opposite sides are always parallel.

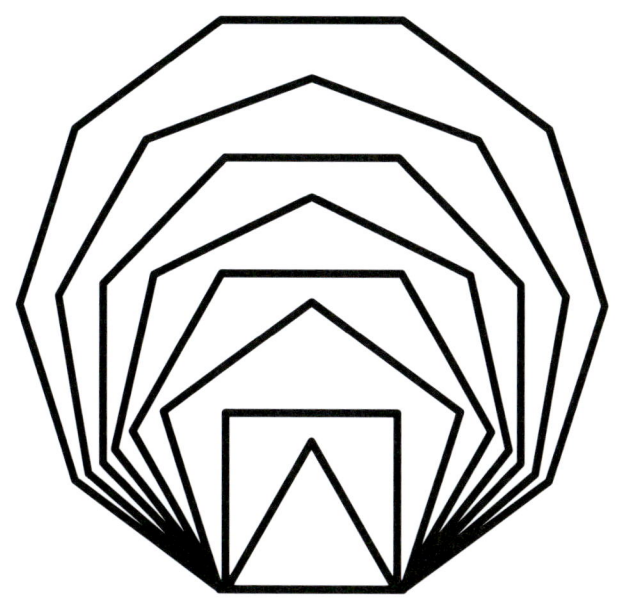

Chapter 3
Symmetry

Thinking Corner

> Can you find a line of symmetry for all 3 shapes in the pattern?
>
> Are there more lines of symmetry for any of them?
>
> Is there a link between the number of triangles in the pattern and the lines of symmetry?

Fact:

Isosceles means 'equal legged'. The name describes the triangle property!

Thinking Corner

How many lines of symmetry are there in each of these shapes?

Count up the number of sides and the number of lines of symmetry — is there a pattern?

Is there a triangle with more lines of symmetry than this one?

Fact:

A square can be called lots of different names. It's also a special type of rectangle, parallelogram, trapezium and more!

Thinking Corner

Can you find any lines of symmetry in the shapes?

Which orientation is easiest to spot any symmetry?

Can you design your own symmetrical shapes using linked squares?

Fact:

Changing the orientation of a shape doesn't change how many lines of symmetry it has, but it can make it easier to find them!

Thinking Corner

Are there lines of symmetry in all of these triangles?

Can a scalene triangle have lines of symmetry?

Can an isosceles triangle have no lines of symmetry?

Fact:

If a shape has no lines of symmetry, we can say it is asymmetrical.

Thinking Corner

Which of these shapes has a line of symmetry?

Do any have more than one line of symmetry?

Fact:

Regular shapes have the same number of lines of symmetry as they do sides.

Thinking Corner

> Which shapes are symmetrical?
>
> Can you add a square to any shape to give it more lines of symmetry?
>
> Is it possible to combine these shapes into a rectangle?

Fact:
These shapes are known as tetrominoes as they're made up of four squares each and joined along whole edges.

Thinking Corner

Which of these shapes have a line of symmetry?

If you change the colour of one square on each picture, can you add more lines of symmetry?

Fact:

The word rectangle just means 'right angle'.

Thinking Corner

Can you find a line of symmetry for each shape?

Do any have more than one line of symmetry?

Can you add small circles to any image to give it more lines of symmetry?

Fact:

A circle has infinite symmetry – draw any diameter and hey presto!

Thinking Corner

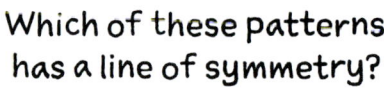

Which of these patterns has a line of symmetry?

Do any have more than one?

Fact:

Shapes made from five identical squares that are joined along whole edges are known as pentominoes.

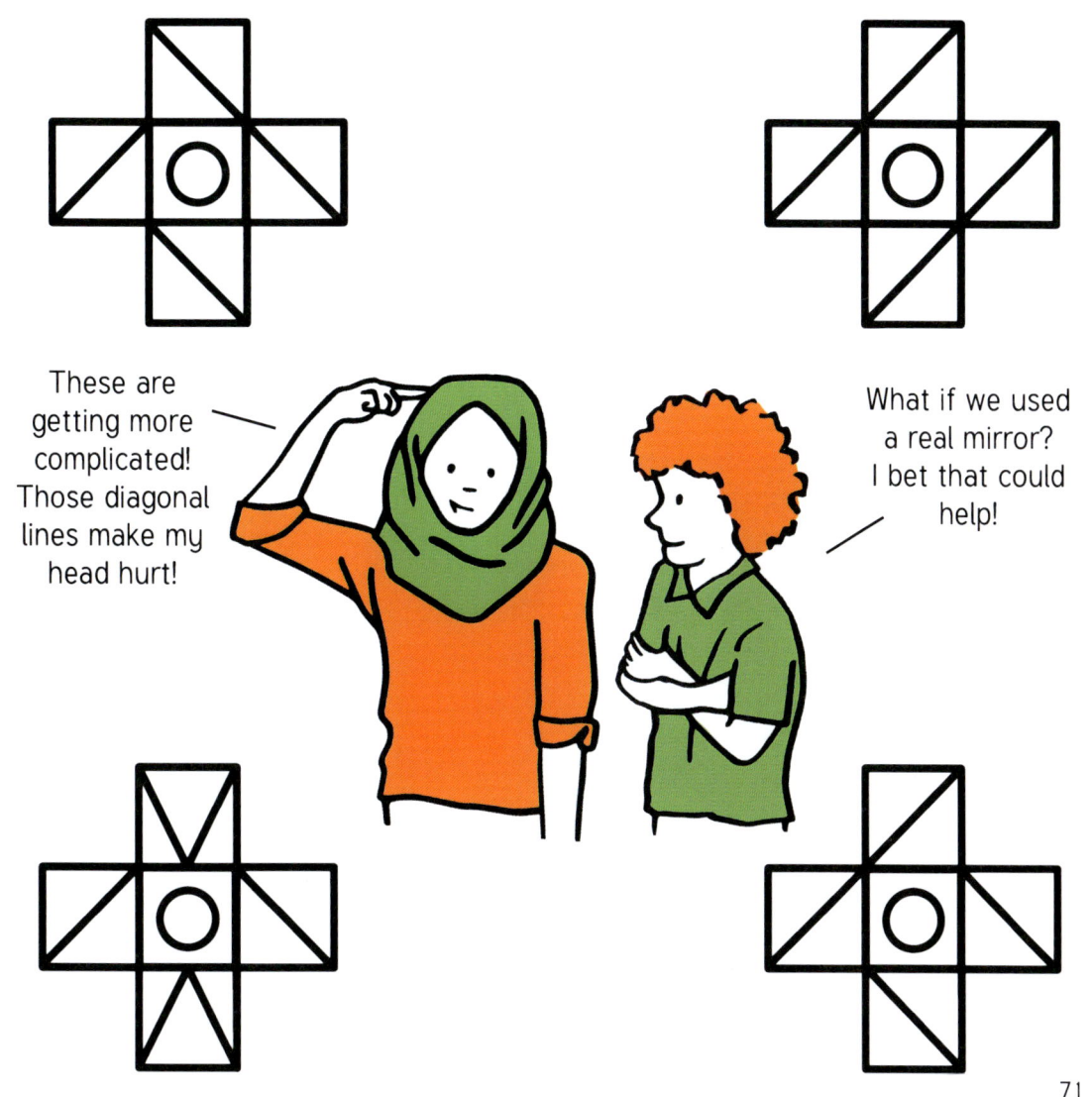

Thinking Corner

Can you add a second circle to each figure to make it have a different line of symmetry?

Can you make a total of four circles on each figure to make them have a different line of symmetry?

Can you make your own patterns on a square grid with different symmetry? Try it!

Fact:

There are different kinds of symmetry. We have looked at lines of symmetry, but there is another kind called rotational symmetry.

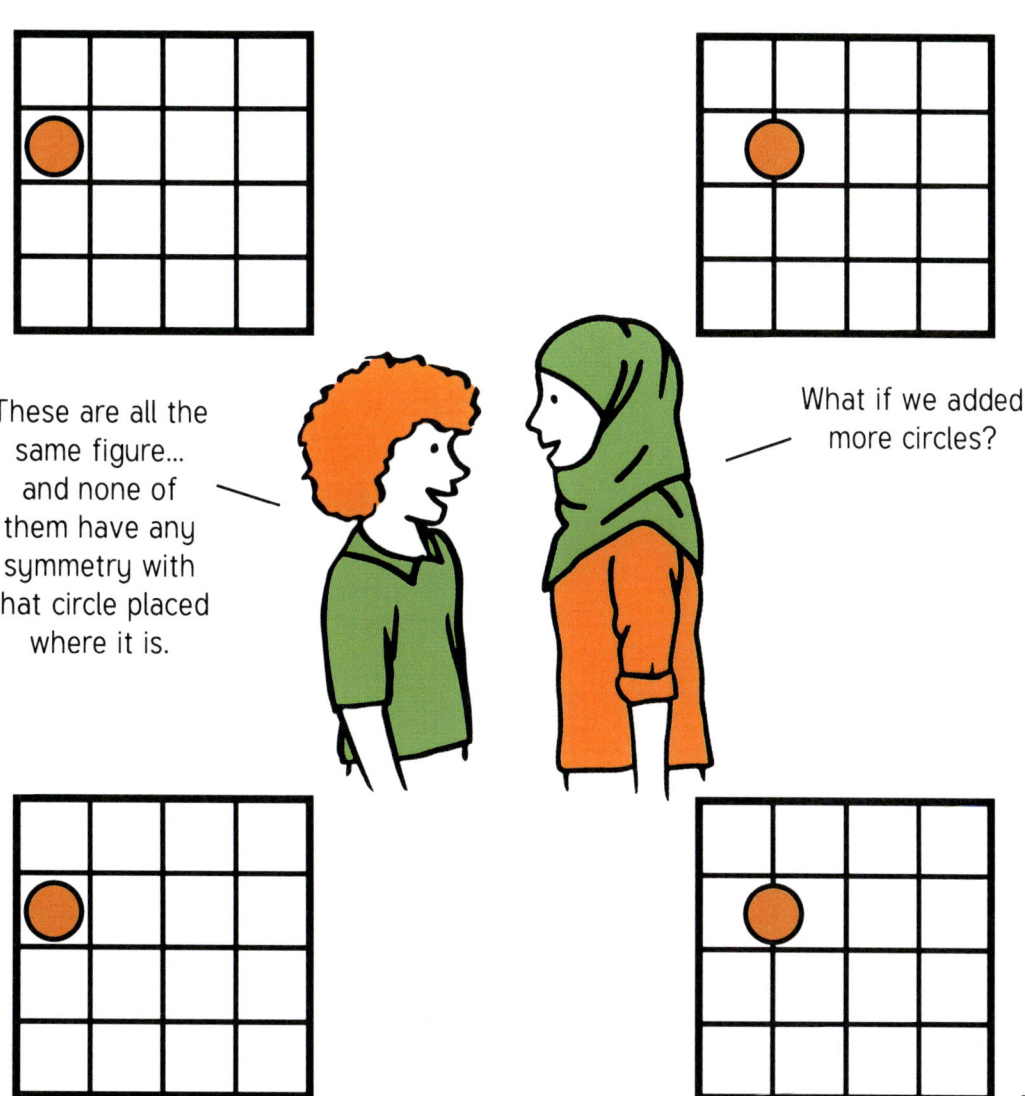

Chapter 4
Cubes

This is the beginning of a pattern. How many cubes would make up the 5th shape?

Can you draw the resulting cuboid?

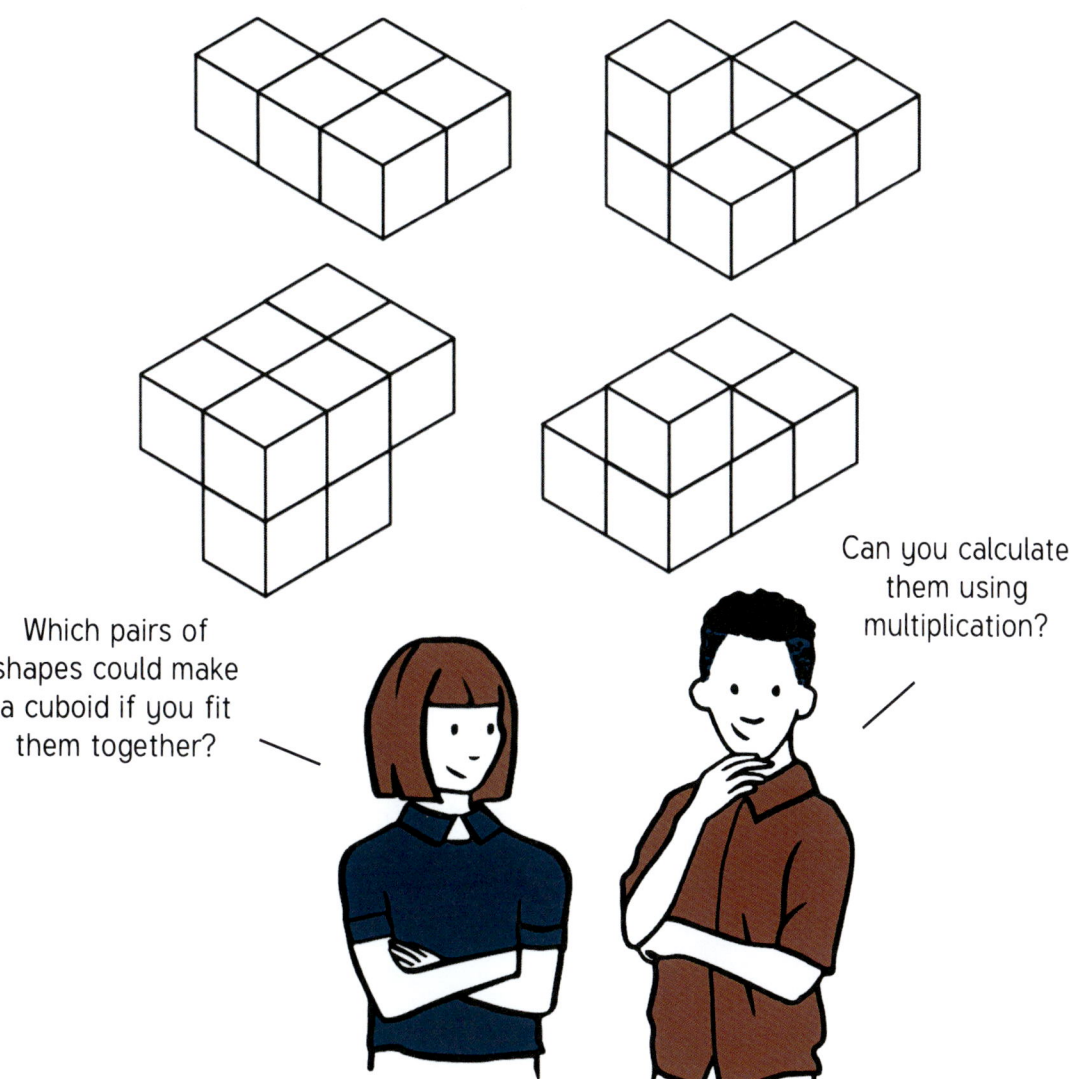

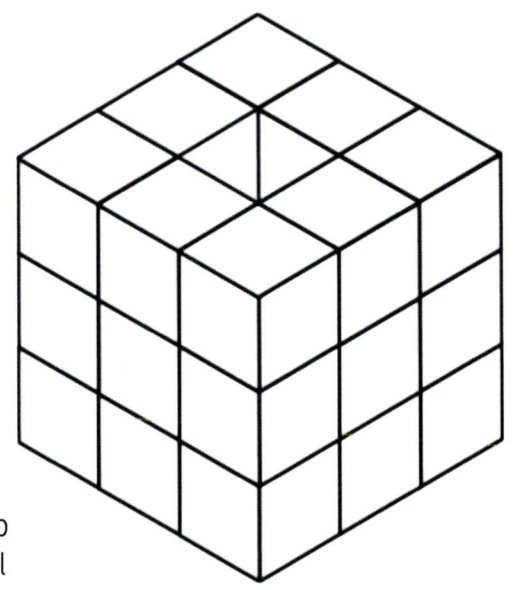

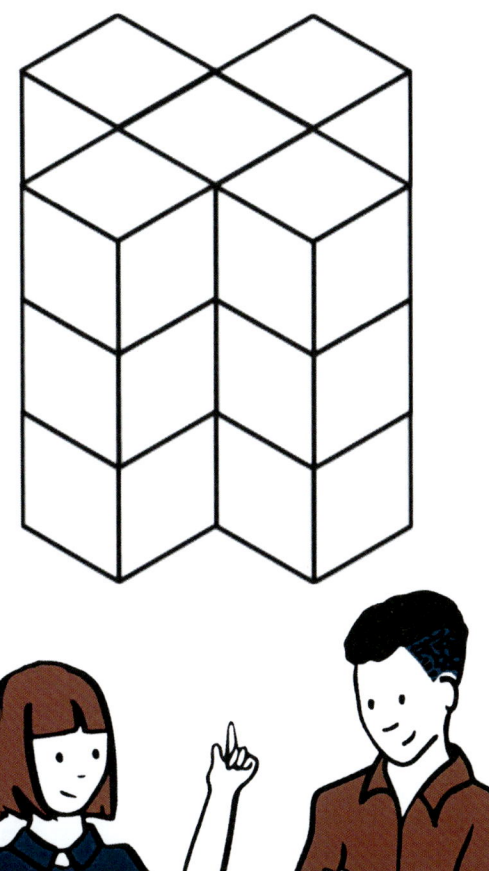

How many more cubes need to be added to this shape to make a 3 x 3 x 3 cube?

If we broke this shape apart, could it be rebuilt into a single cube shape?

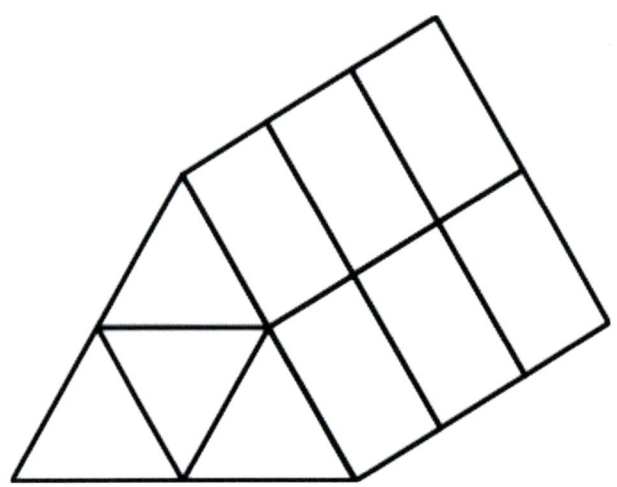

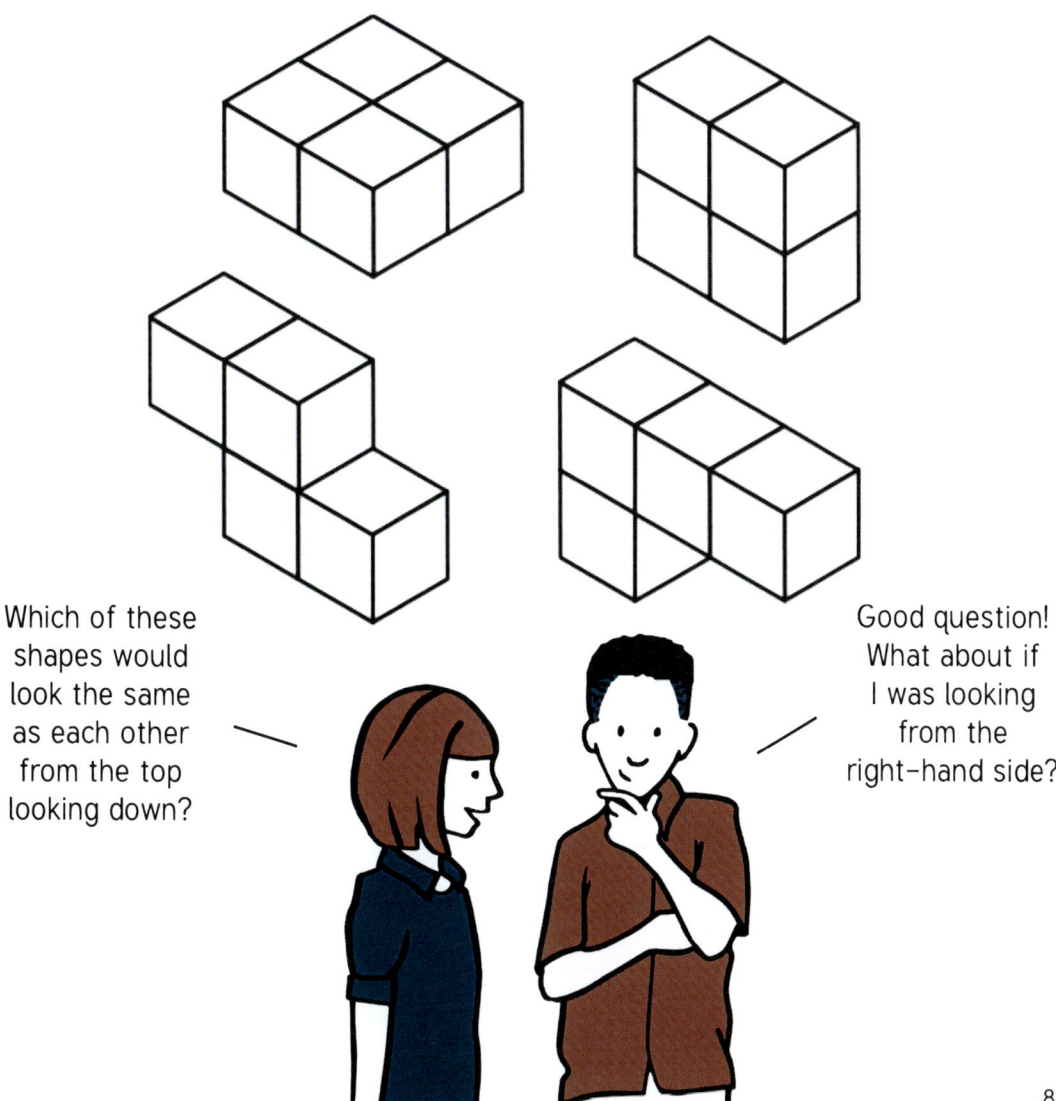

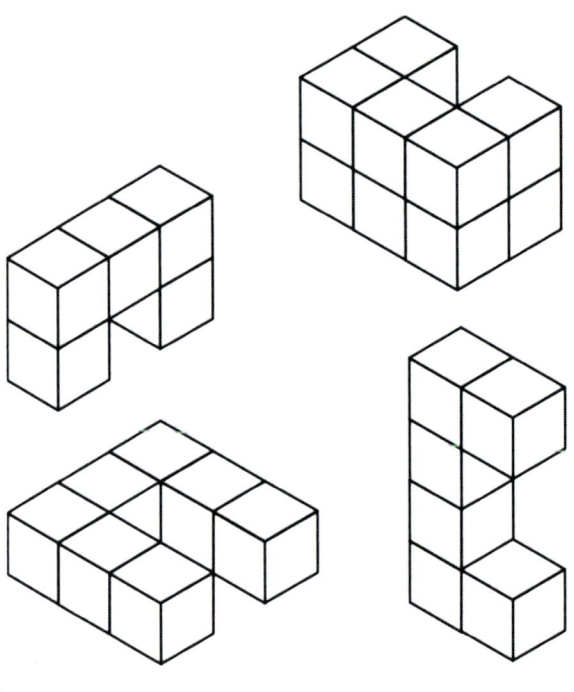

What is the tallest shape you could make by balancing these four shapes on top of each other?

Can any of these shapes be reassembled into a cuboid in more than one way?

Chapter 5
Coordinates

Thinking Corner

> What would the coordinates be for a fourth point to make a square?
>
> Can you place a fourth point to make a different triangle?

Fact:

Using coordinates in maths was popularised by mathematician René Descartes in the 1600s, which is why they're often called Cartesian coordinates.

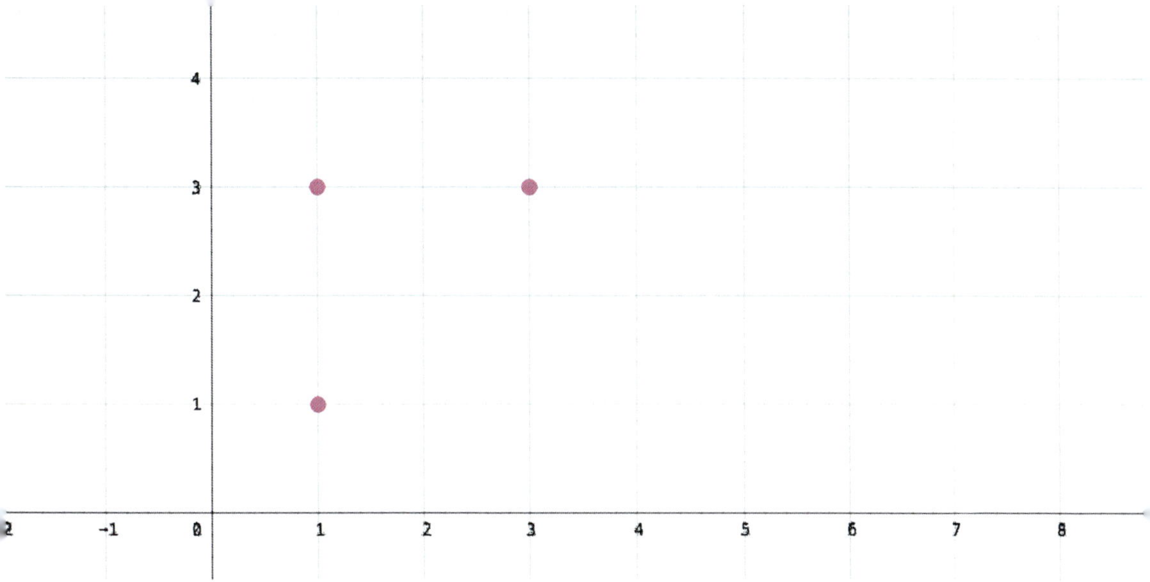

Thinking Corner

Can you add a further two points to make a square?

Can it be done another way?

Fact:

Diagonal includes 'gon' like with shape names. Diagonal means 'from angle to angle'.

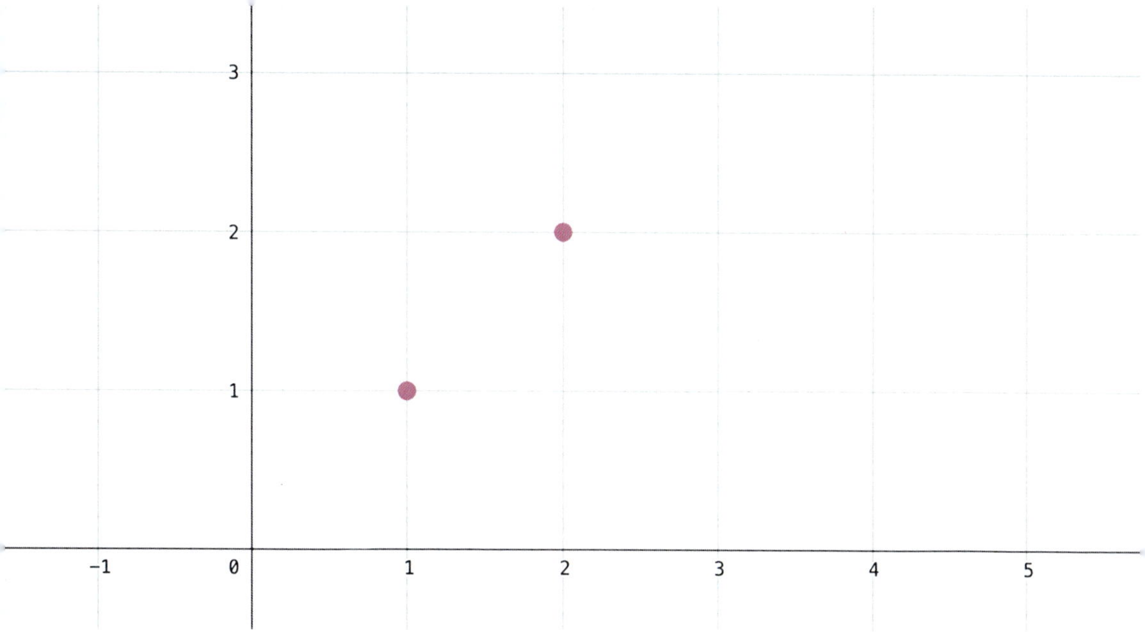

Thinking Corner

Can you make a parallelogram by adding one more point?

Will one more point always make a quadrilateral?

Fact:

We're using positive coordinates in this example, but you can use negative coordinates in other **quadrants** too!

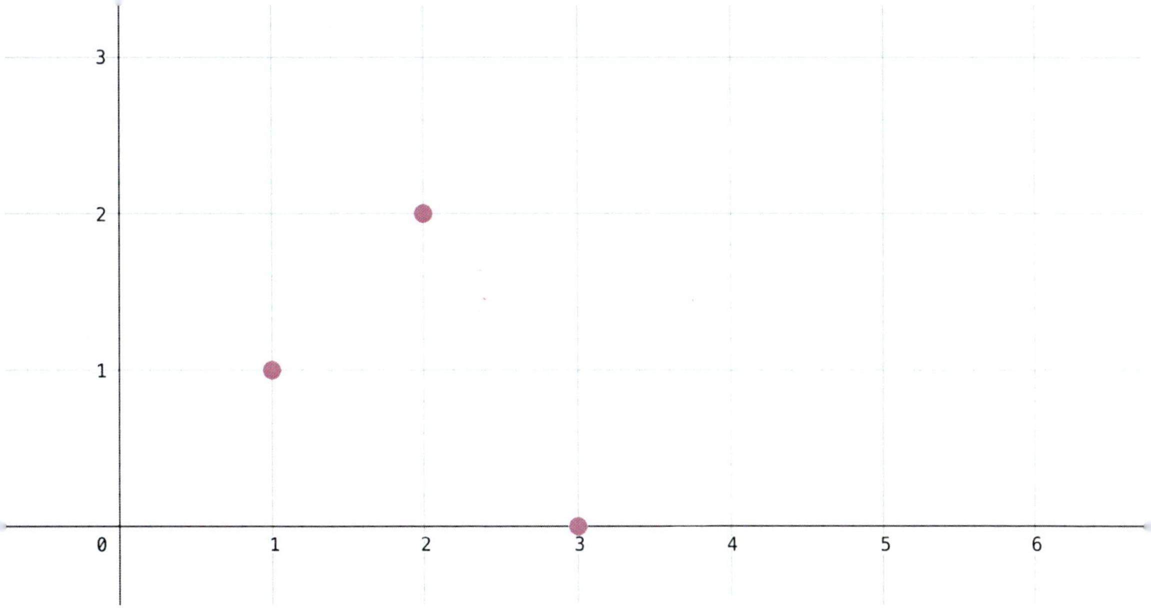

Thinking Corner

Can you make a trapezium with these points?

Can you add a point to make a parallelogram?

Can you add a point to make a right-angled triangle?

Fact:

The word 'parallel' is often misspelt in maths. An easy way to remember how to spell it is that it has two parallel lines within the spelling, the 'l's!

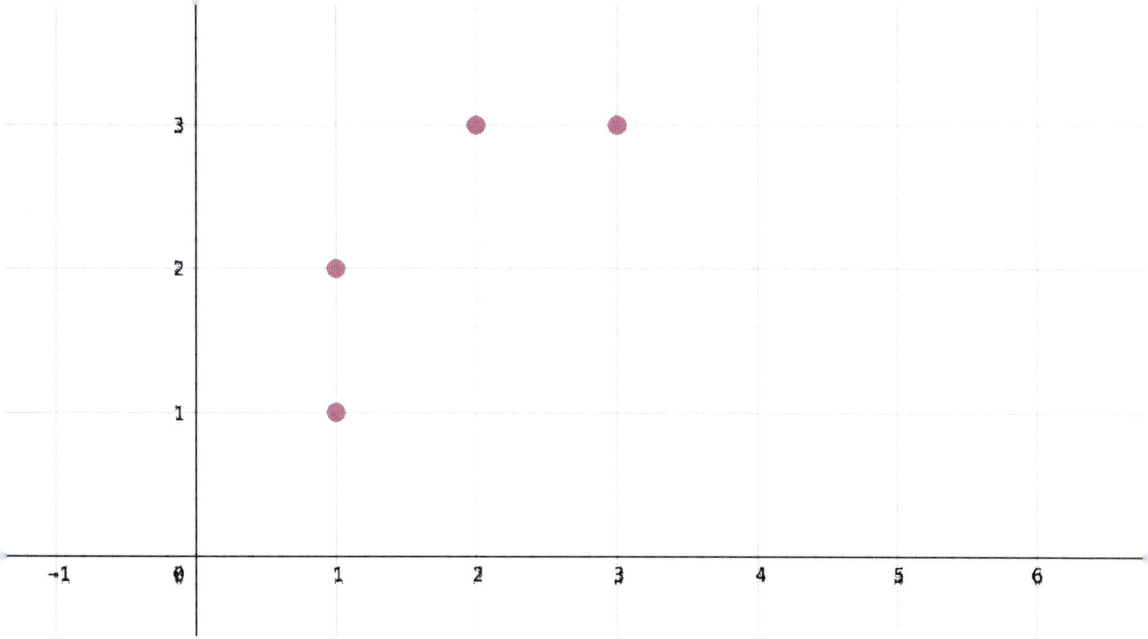

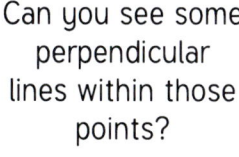

"Can you see some perpendicular lines within those points?"

"No, maybe you can help me? I *can* see parallel lines though!"

Thinking Corner

What are the coordinates of each vertex?

Can you redraw this square four units to the left?

What are the coordinates of the points now?

Fact:

Moving shapes on a coordinate grid without changing the size is called a **translation**.

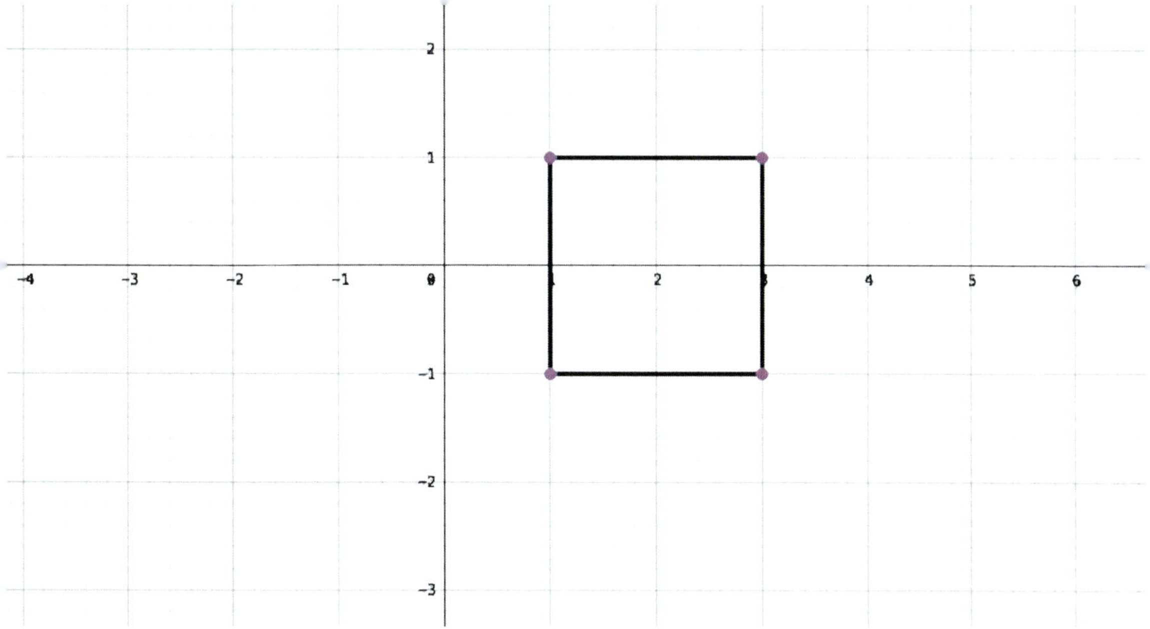

Thinking Corner

What's the area of the square?

Can you draw another square the same distance away from the point (2,2)?

Visualise a square twice as long. Is the perimeter twice as big?

What about the area?

Fact:

In maths, if we change the size of a shape, but keep it all in proportion, we call it a **similar shape**.

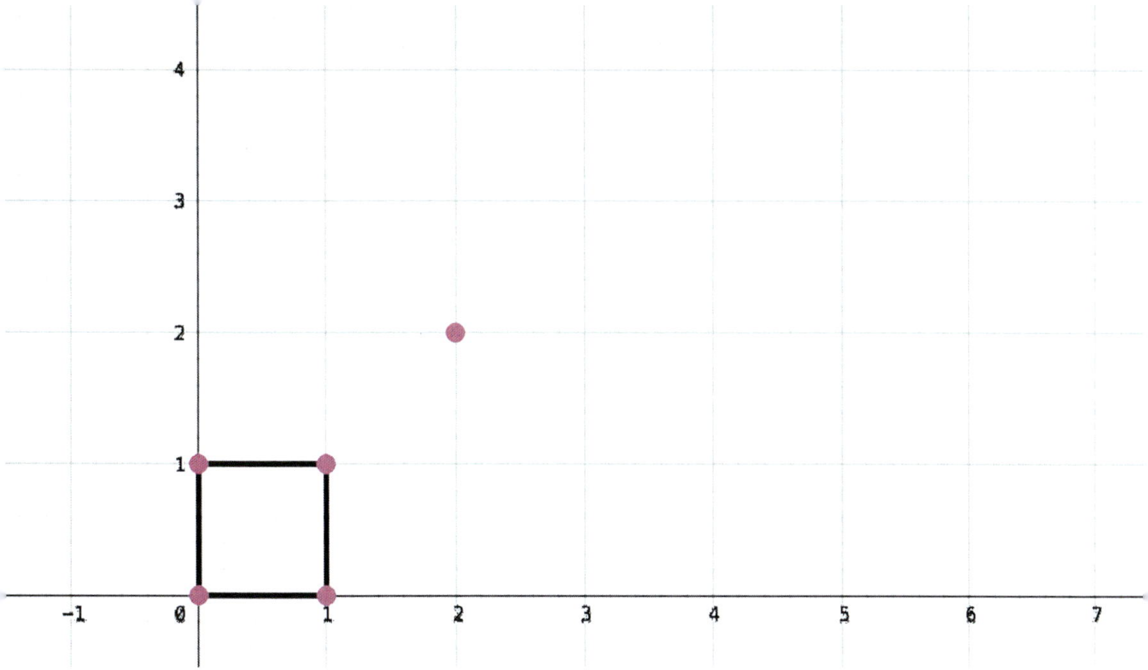

Thinking Corner

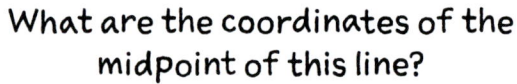

What are the coordinates of the midpoint of this line?

What if we split the line into thirds?

Where would we mark it?

Can you draw a line perpendicular to this one?

Fact:

We can describe straight lines plotted on a pair of axes using an **equation**.

Thinking Corner

What are the coordinates of the three points?

What would the reflected triangle of points look like?

What would the coordinates of the reflected dots be?

Fact:

The word 'reflection' is related to the word 'reflex' which means "bent back."

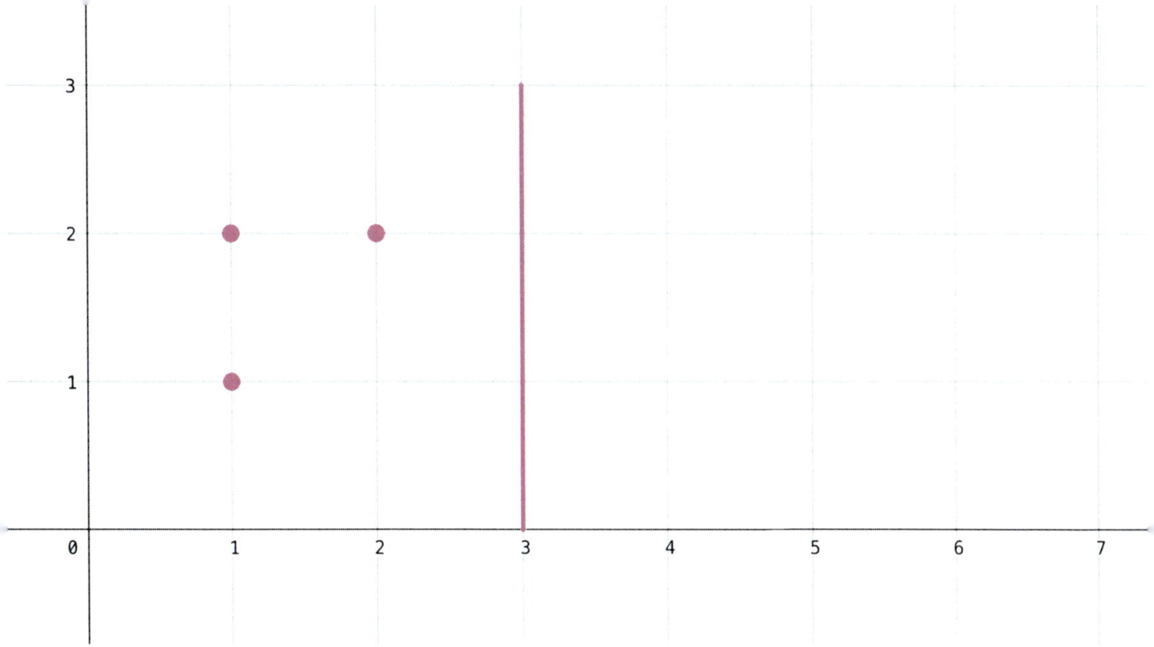

Time for a challenge! What if we reflect these points across a mirror line starting at (3,0)?

Does that mean the reflected shape would look exactly the same?

Thinking Corner

Can you complete the pattern to include a fourth triangle?

What shape is made by the shortest side of each triangle?

Is there a way to find the area of any of the triangles?

Fact:

A regular four pointed star is called an isotoxal star.

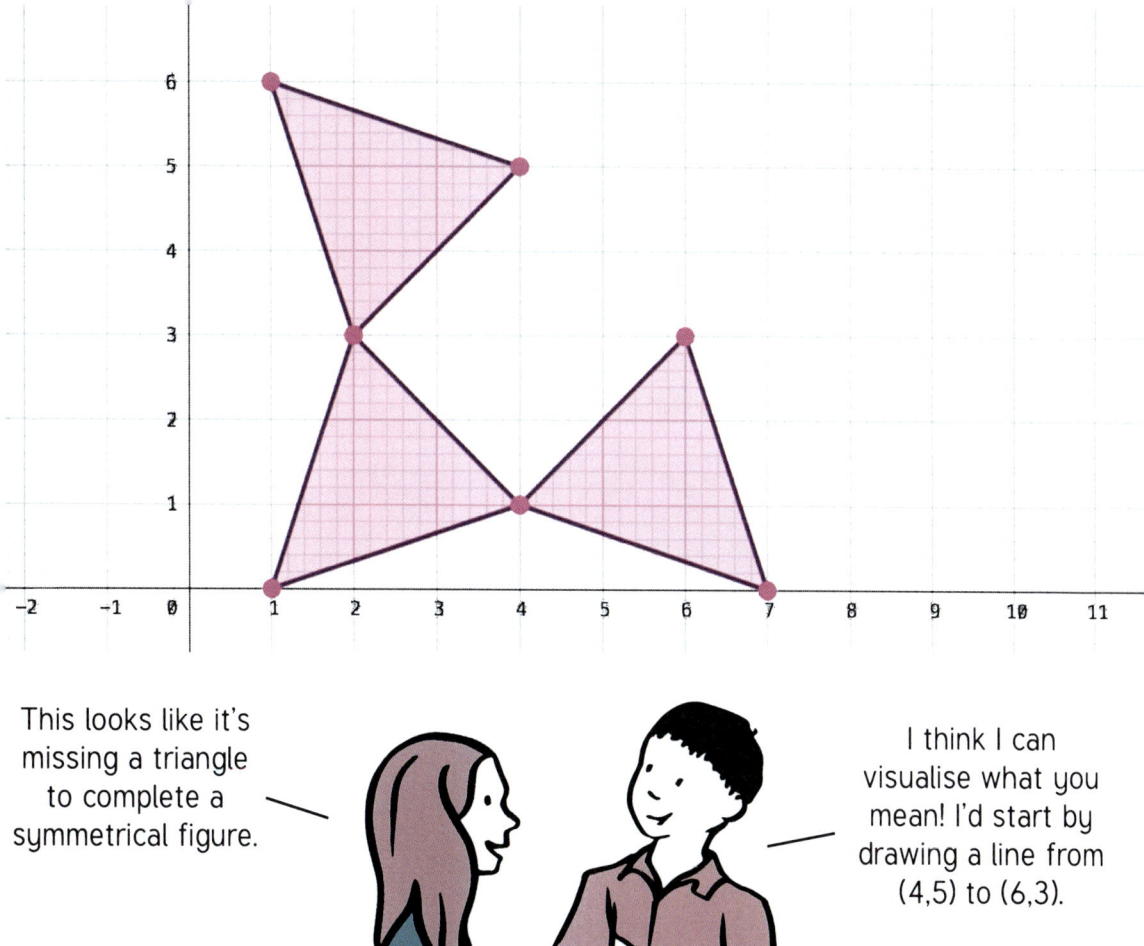

Thinking Corner

Following the pattern to the right, where would the next triangle go?

What would the next coordinate be?

What would the coordinates be for the eighth triangle in the sequence?

Fact:

Tessellations are where shapes fit together continuously leaving no gaps between them.

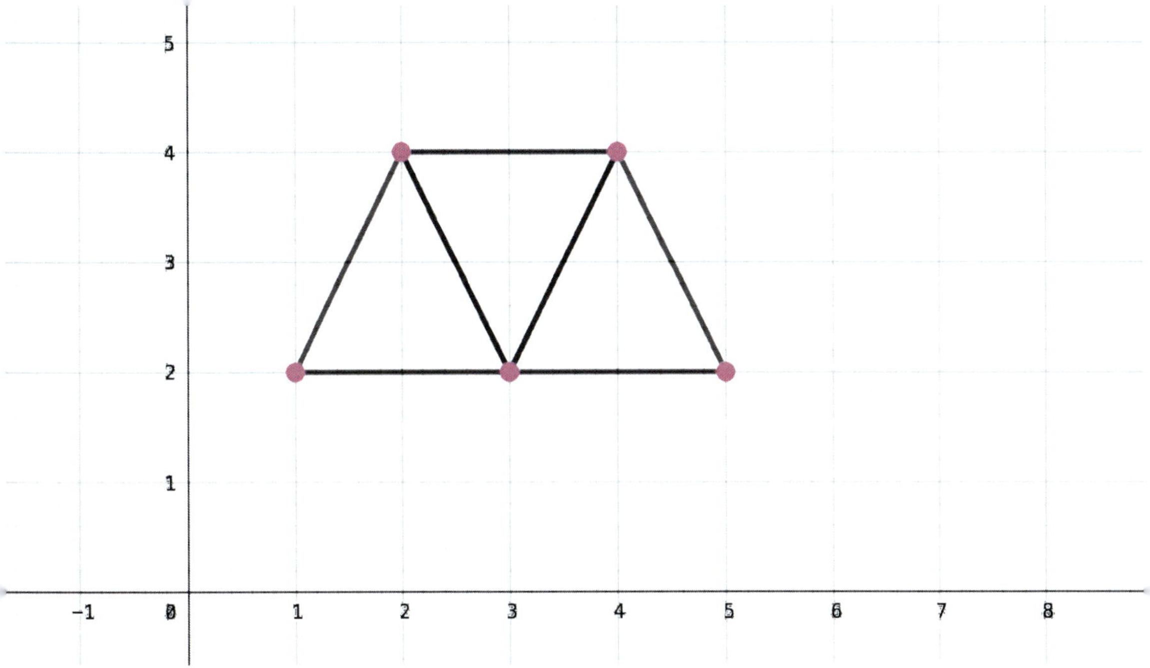

Chapter 6
Nets

Thinking Corner

> Which nets make cubes?
>
> Can you think of any different cube nets?
>
> How could we make sure the volume of each cube is the same?

Fact:

The term 'net' in maths can be confusing because of the use of the word in fishing. It might be easier to think of it in terms of 'spread' or 'bound together'.

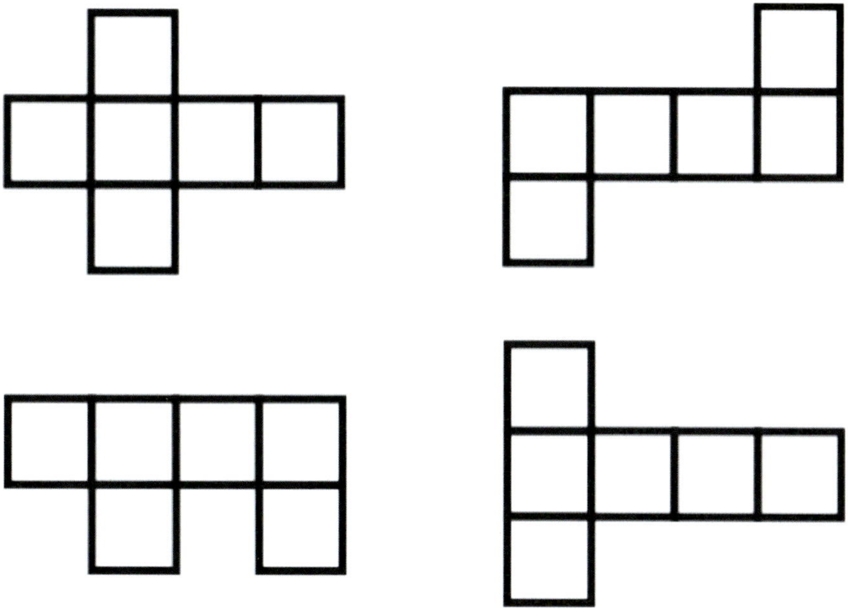

Thinking Corner

What 3D shape is made up of a square face and four identical triangular faces?

Do all these nets make that shape?

How can you be sure?

Fact:

A 3D shape with five faces can also be called a pentahedron.

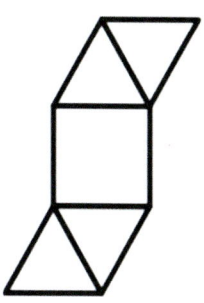 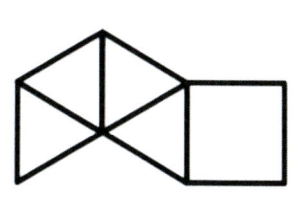  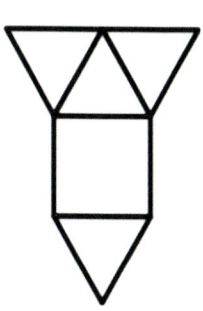

Well I'm certain these won't make cubes!

I agree, but the shape does have one square face. Maybe it's related?

Thinking Corner

Which two shapes have the same number of edges, faces and vertices?

Which of these shapes has the most faces?

Can you draw the net for any of these 3D shapes?

Fact:

A cuboid can also be called a rectangular parallelepiped which is a bit of a mouthful!

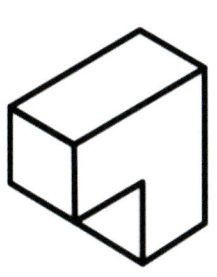 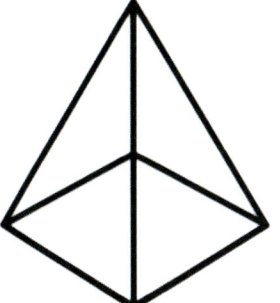 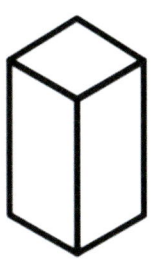 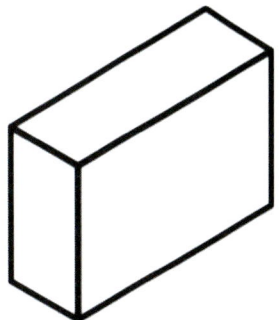

I imagine some of the nets for these shapes will be complicated.

Yes, it gets hard to picture them when a shape has lots of faces.

Thinking Corner

Which of these 3D shapes would have nets that include squares?

Which shape would probably have the longest net? Why?

Can cuboids be made from different types of net?

Fact:

A cuboid has six faces, with each pair of opposite faces being identical rectangles. If you think about it, a cube is a special kind of cuboid!

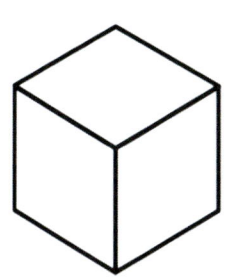 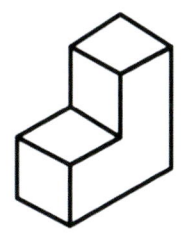 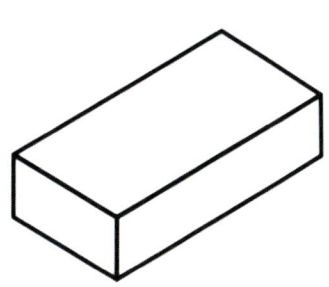 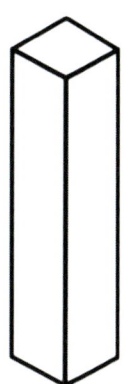

Three of these shapes share very similar properties. Does that mean their nets will look the same?

I think it will mean their nets will have the same structure, but maybe some parts will be different sizes.

Thinking Corner

Which nets will include both squares and rectangles?

Do any of these shapes have triangles in their nets?

If you fit any of these shapes with a copy of itself, what shape can you make?

Fact:

All of these shapes are examples of a family called prisms.

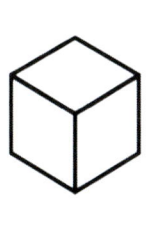 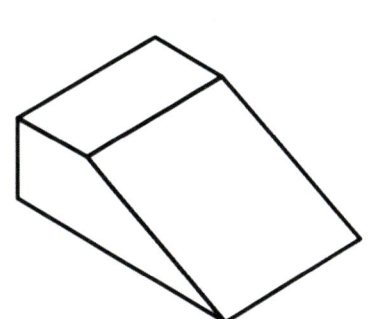 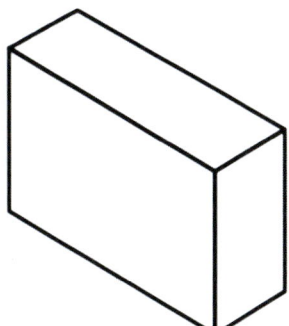 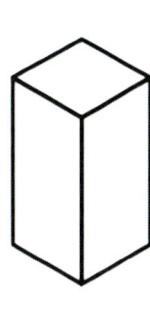

One of these shapes stands out to me. It looks different from the rest.

It looks like a triangle, but if I look closer... I'm not sure if there are any triangles at all?

Thinking Corner

Which shapes have rectangles in their nets?

What would the net of the semi-cylinder look like?

How many triangular faces does each shape have?

Fact:

A shape is known as a prism if its end faces are the same, and the other faces are rectangles. Mathematicians sometimes disagree about whether a cylinder is an example of a prism.

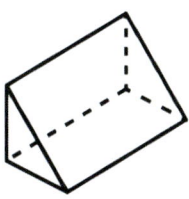 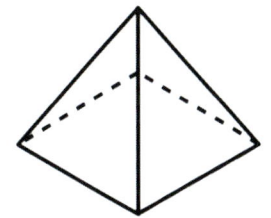 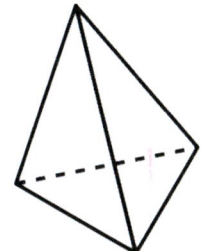 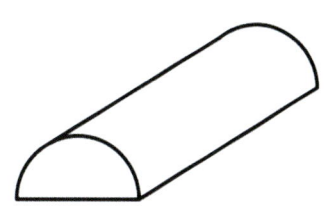

These shapes are interesting! I don't see how the curved one is related to the rest.

I thought that at first, but look closer – the other three aren't *all* just made of triangles...

Thinking Corner

What will these nets look like in 3D?

How many edges will each 3D shape have?

How many vertices?

Fact:

There is a special relationship between the number of edges (E), faces (F) and vertices (V) in 3D shapes called Euler's Formula: $F + V - E = 2$.

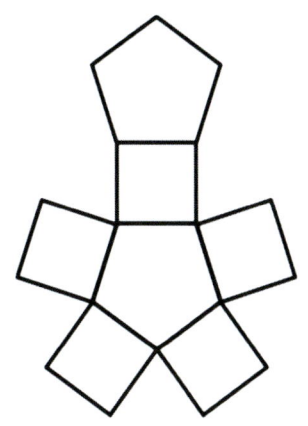
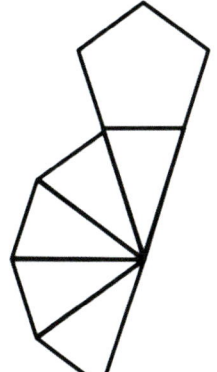

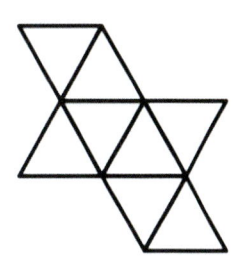

These shapes look complicated!

If you pick the biggest shape and imagine it's the 'bottom', I think it's a bit easier to figure out.

Thinking Corner

Have a look at a real pair of dice. What do you notice about opposite sides?

Which of these nets would make 'real' dice?

What are the chances of rolling a number greater than four on a die?

Fact:

The earliest dated dice that archaeologists have discovered date back to around 2800 BC.

Thinking Corner

Which cubes will form 2 closed loops with their patterns?

Could you tessellate any of these nets?

If the squares had side length 2cm, what would the volume be for each cube?

Fact:

When we talk about 'cube numbers', their name is related to the dimensions of a cube. For example, 27 is a cube number, which can be represented as a 3 x 3 x 3 cube.

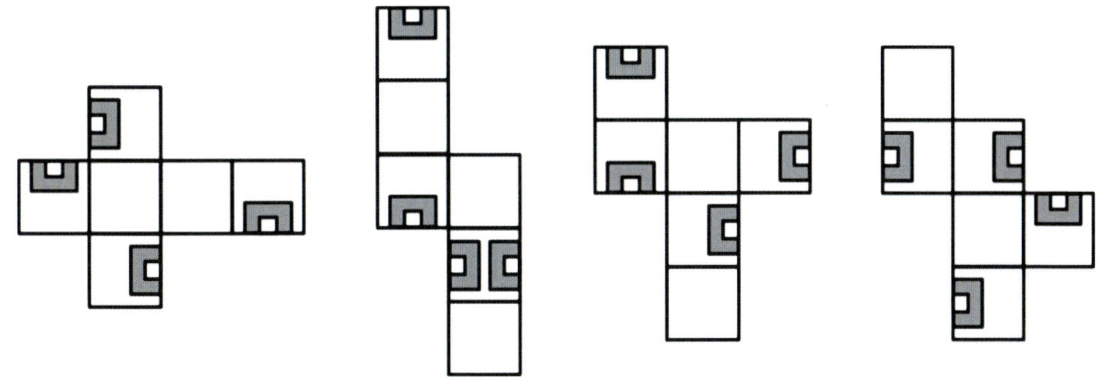

These cube nets have interesting patterns on them.

If they were folded into cubes, would the patterns all look the same?

Thinking Corner

Can you sketch the net for this shape?

How many edges does this shape have?

If this shape is made from 3 cubes, how many more would it take to turn it into a bigger cube?

Fact:

It may be tempting to think that this shape has ten vertices, but it does in fact have twelve. The two 'inward facing' (convex) vertices are still places where edges meet.

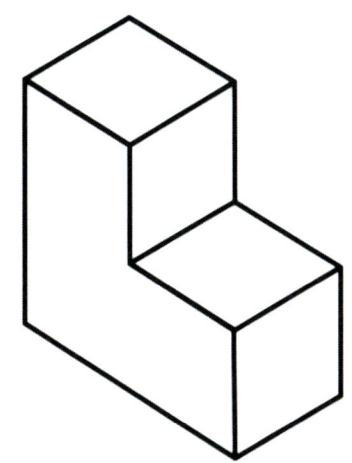

I see this as a cuboid with a bite taken out of it!

Ooh, I saw it more as an L shape, but you're right, and the 'bite' looks like a cube.

Chapter 7
Angles

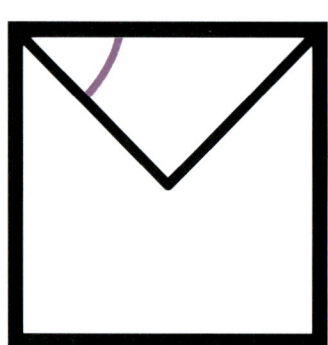

Let's test your skills! Here we have a triangle inside a square with one vertex in the middle. See what angles you can figure out! As a challenge, try the one I've highlighted.

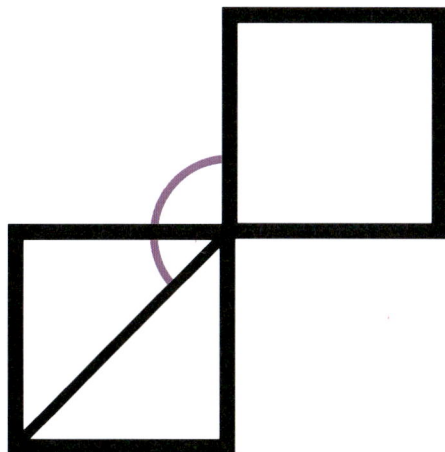

This time we have two identical squares either side of a vertical line. What can you tell me about the different angles here? Can you find the one I've highlighted?

Another square, but this time it has an equilateral triangle on top. There are lots of angles here. See which ones you can figure out. You don't have to try the one I've highlighted straight away, it might be easier to play around first.

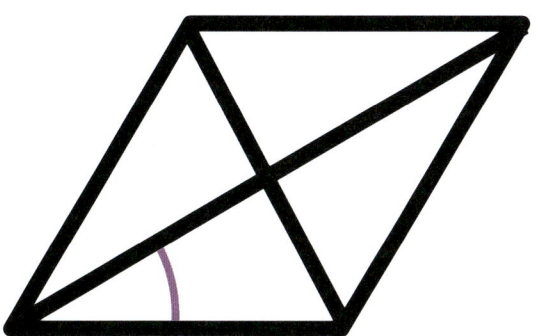

No squares this time! Just two equilateral triangles sharing a side. There are more triangles in this picture to think about though. Maybe those can help you find the angle I've highlighted?

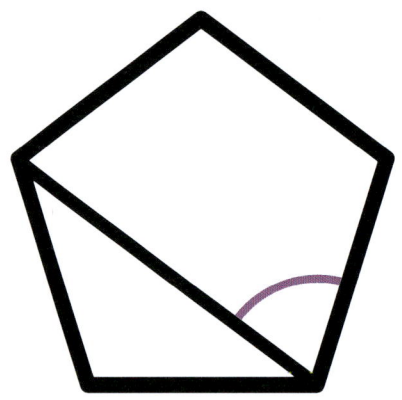

OK here's a more complicated shape: A regular pentagon. It's split into two other shapes though. Do you know their names? If we use properties of all these shapes we might be able to find the highlighted angle?

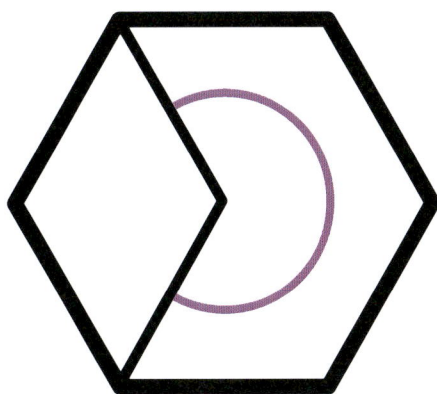

Less help this time! Here's a regular hexagon, and the lines inside it meet at the middle. Can you find the highlighted angle?

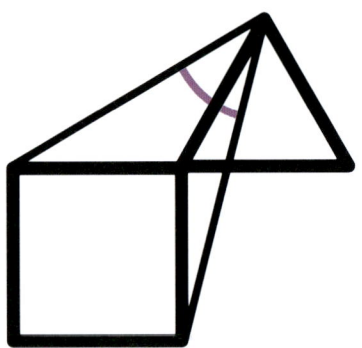

This time I'm going to give you an **assumption**. If we assume the square and equilateral triangle have the same length sides, and they sit below and above a horizontal line, can you find any angles in the picture?
As a challenge, try the one highlighted.

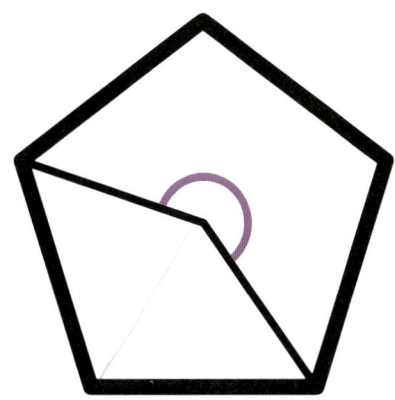

Back to the pentagon! It has two lines coming out of the centre going to different vertices. Can you find the value of the highlighted angle?

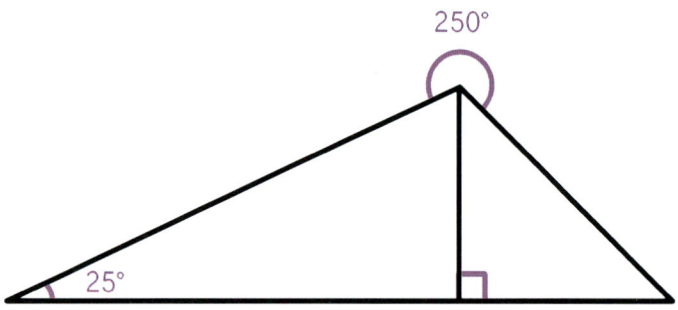

This time I've given you some angles as a head start. Can you find all the angles I haven't given you?

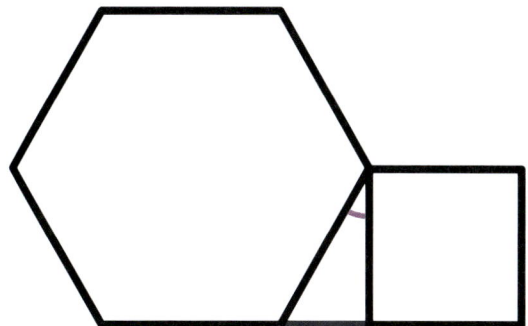

You're doing so well! OK let's see if you can handle a regular hexagon and a square together on a straight horizontal line. They have one touching vertex. Can you tell me any of the angles in this picture? What about the one I've highlighted?

Chapter 8

Areas

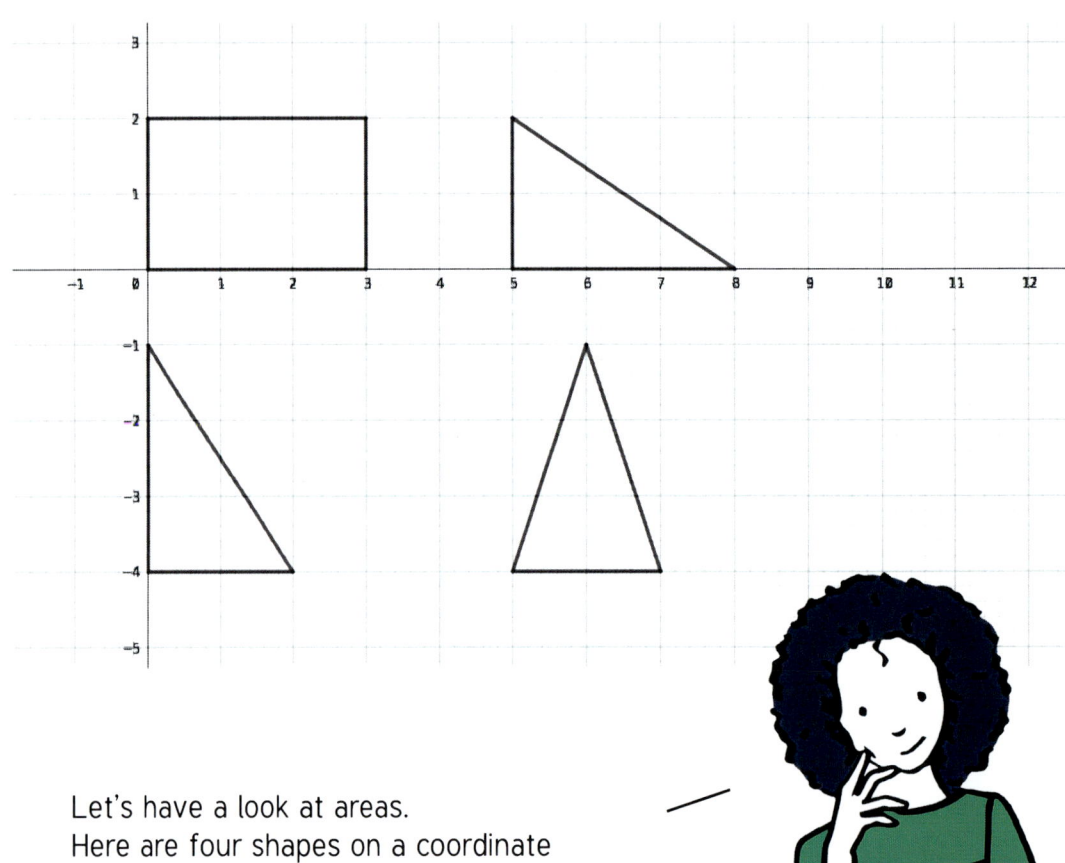

Let's have a look at areas.
Here are four shapes on a coordinate grid. What is the area of each shape?
Do you spot anything interesting? What is it?

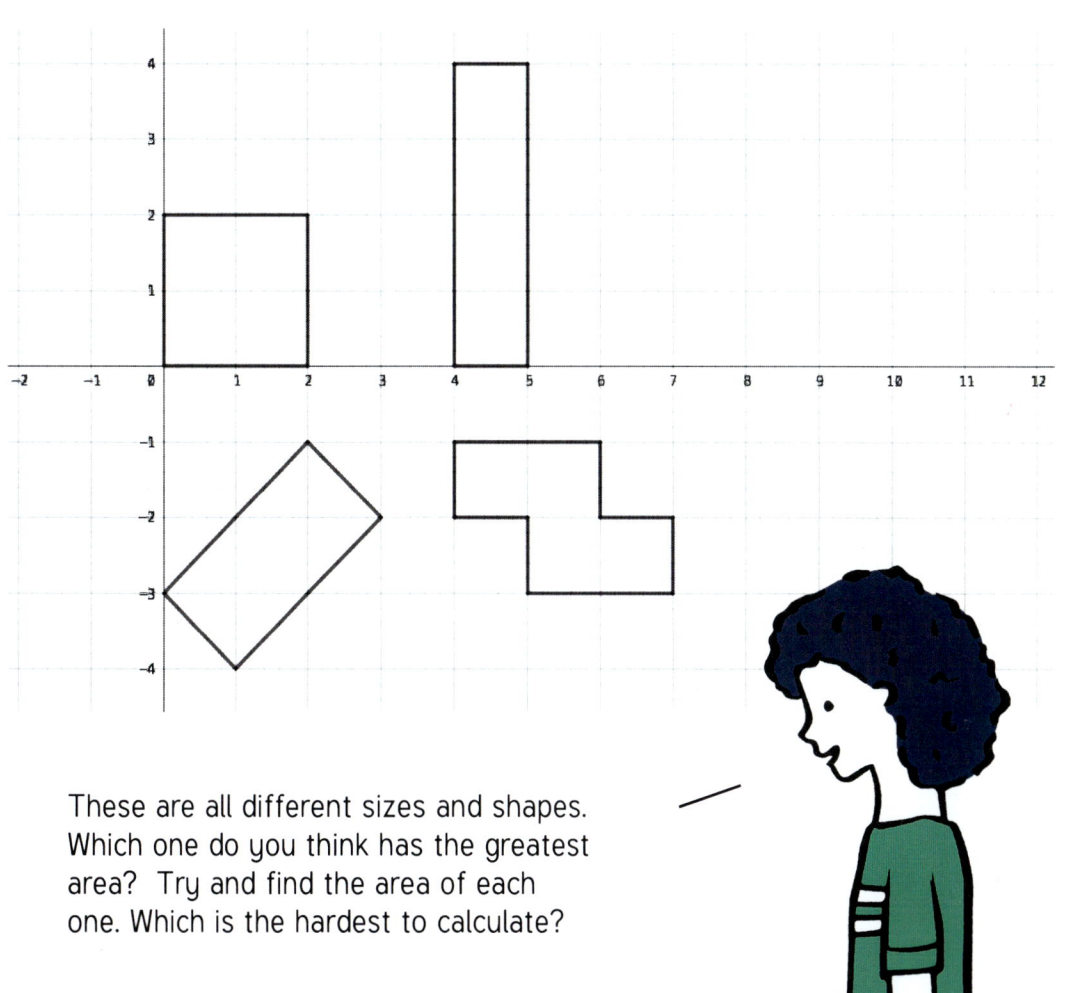

These are all different sizes and shapes. Which one do you think has the greatest area? Try and find the area of each one. Which is the hardest to calculate?

No grid this time! What do you notice about the shaded square in each picture? Does the area look the same each time or different? What about the white part of the bigger squares? Estimate what proportion of each square is shaded!

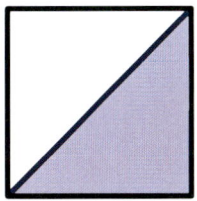

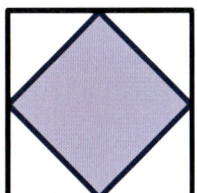
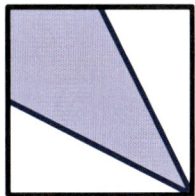

Here are four squares with equal area. The shaded regions are made using corners and *midpoints*. What can you tell me if we compare the area of the shaded regions? If each large square had a side length of 1, what would the shaded areas be?

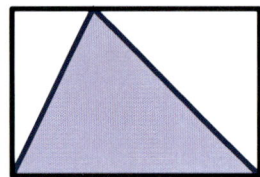 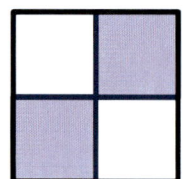

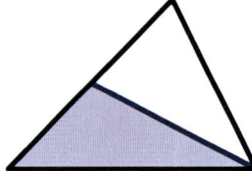 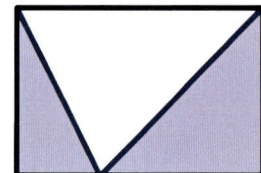

What about these shapes?
If each one had the same area, what would the area of the shaded sections be?
Do you need more information?

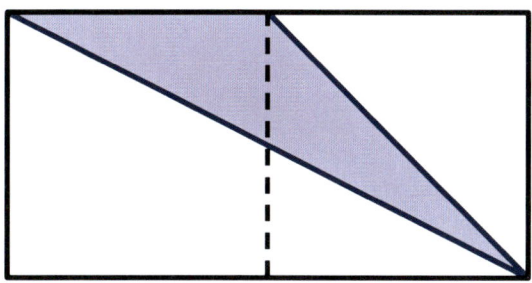

This rectangle is made of two squares. If the area of the shaded triangle is 8cm², what are the areas of each white triangle?

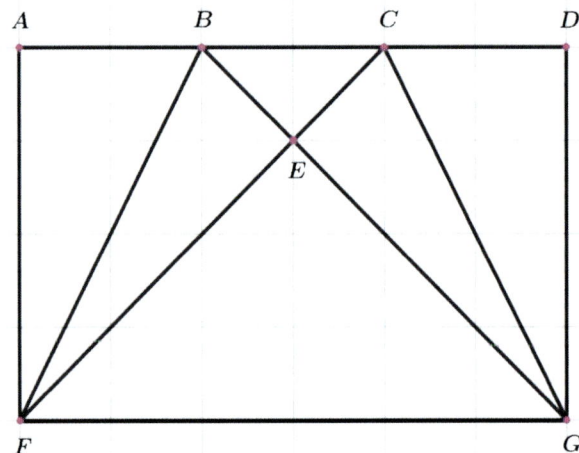

Using three letters to describe the triangles (for example "triangle FBA") you're looking at, can you find any pairs that have the same area? How do you know?

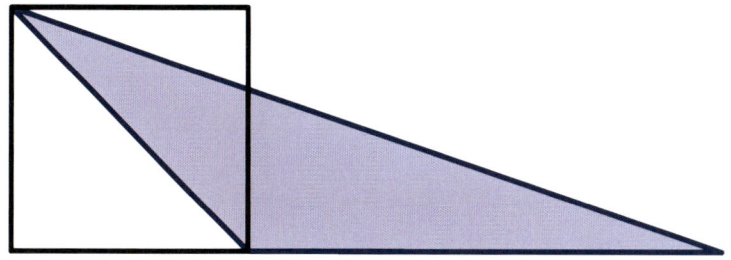

If this square and large shaded triangle have the same area, how much longer is the base of the triangle compared to the square? Assume they both sit on a horizontal line.

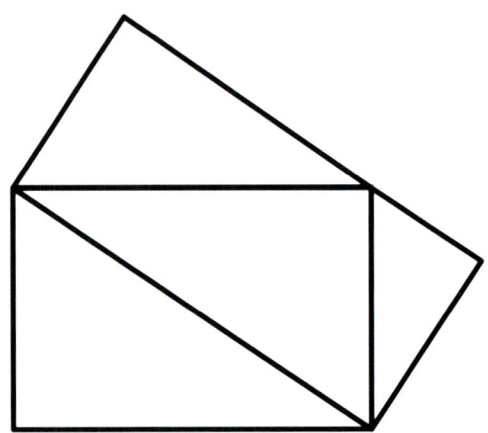

This one might take some extra thinking time! What is the relationship between the areas of the two rectangles?

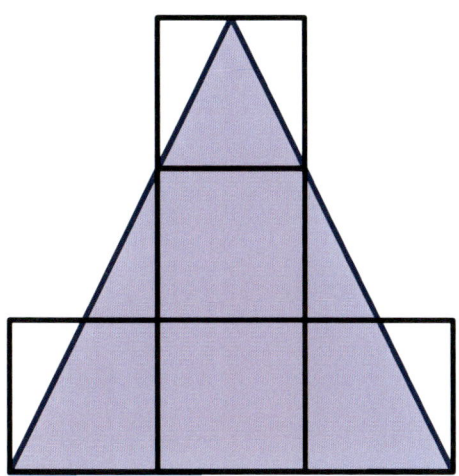

If we say each square has an area of 1cm², can you figure out the area of the shaded triangle? It might help to draw some more guidelines!

Chapter Notes

Shape Definitions:

Regular shapes: *A polygon with equal sides and equal angles.*

Irregular shapes: *A polygon that is not regular.*

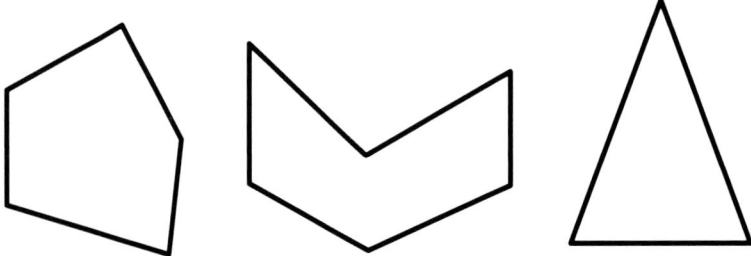

Triangles:

An **equilateral** triangle is a regular triangle – all sides are the same length.

An **isosceles** triangle has two sides of identical length.

A **scalene** triangle has no sides of equal length.

A **right-angled** triangle has an internal angle of 90°.

Quadrilaterals:

A **parallelogram** is a four-sided shape with parallel opposite sides.

A **rhombus** is an equilateral parallelogram.

A **trapezium** is a four-sided shape with at least one pair of parallel sides.

A **kite** is a four-sided shape with two pairs of adjacent sides of equal length and perpendicular diagonals.

A **rectangle** is a four-sided shape with opposite sides of equal length, and internal angles of 90°.

A **square** is a four-sided shape with equal length sides and internal angles of 90°.

Circles:

A **circle** is a set of points equidistant from a single point (the centre).

Symmetry:

Regular shapes have the same number of lines of symmetry as they do sides.

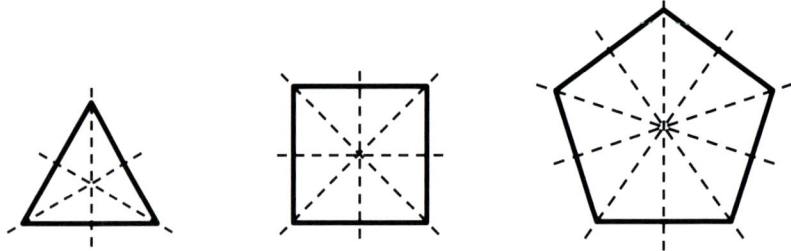

Here are some other examples of lines of symmetry:

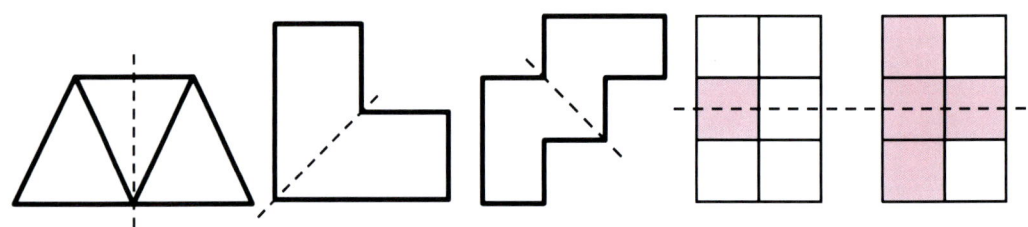

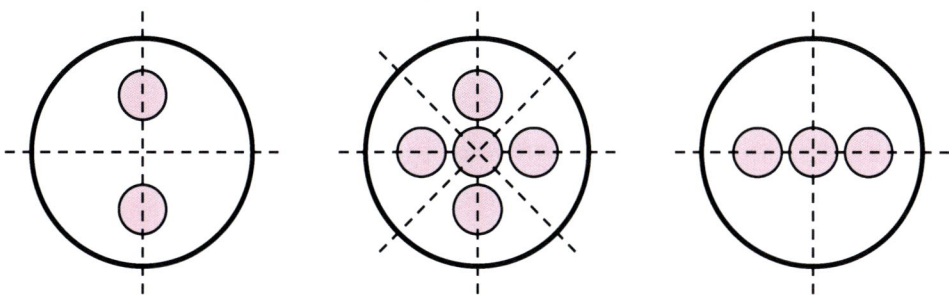

Cube Puzzles:

1. 5 cubes make up the shape, a further 3 would make a 2 x 2 x 2 cube.

2. 5 x 5 x 5 = 125

3. The greatest number of cubes is dependent on how many cubes are hidden at the back. From the angle in the diagram, there could be as many as 24 cubes in total. The smallest number would be 13 which is the number you can see from the diagram.

4. Either piece on the right-hand side would make a cuboid with the piece that is only one cube tall.

5. 9 x 2 + 3 x 2 = 24. If the hole is only one small cube in size, then the volume would be 3 x 3 x 3 − 1 = 26 cubes.

6. 3 x 4 = 12 cubes (assuming the original shape is made of 5 pillars that are 3 cubes tall). If the original shape is made up of 15 cubes, then it cannot be made into a cube as 15 is not a cube number.

7. 4 x 3 = 12 prisms. Each face is either a rectangle or a triangle.

8. If we do not rotate any shapes, then a birds-eye view (a plan view) would be the same for two shapes — a rectangle of 3 squares.

9. All shapes can be rearranged into a cuboid with a single cube width. Two can be arranged another way:

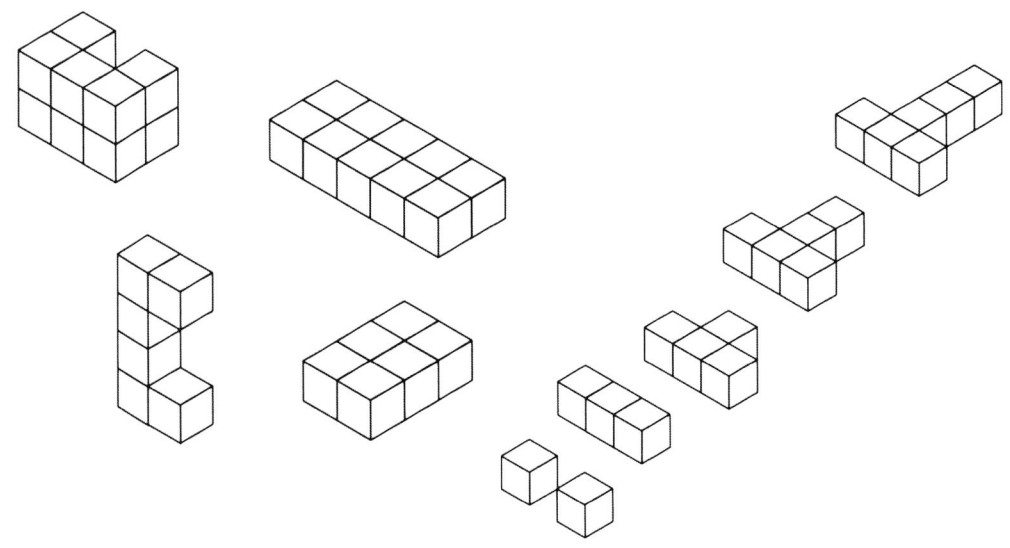

Coordinates

Task 1: A square can be made using (3,1), and different triangles can be made using a further point at (0,3) or (1,4) for example.

Task 2: A square can be made using (1,2) and (2,1). Alternative squares can be made using (0,2) and (1,3) or (2,0) and (3,1).

Task 3: A parallelogram can be made by adding the point (4,1). Adding a point such that three points are on a straight line would mean joining the four points would not make a quadrilateral, for example (0,0) or (3,3).

Task 4: Adding a point at (0,1) or (3,4) can make a parallelogram. Adding a point at (1,3) would make a right-angled triangle.

Task 5: The original square has points on the coordinates (1,−1), (1,1), (3,1) and (3,−1). The translation would have points on the coordinates (−3,−1), (−3,1), (−1,1), (−1,−1).

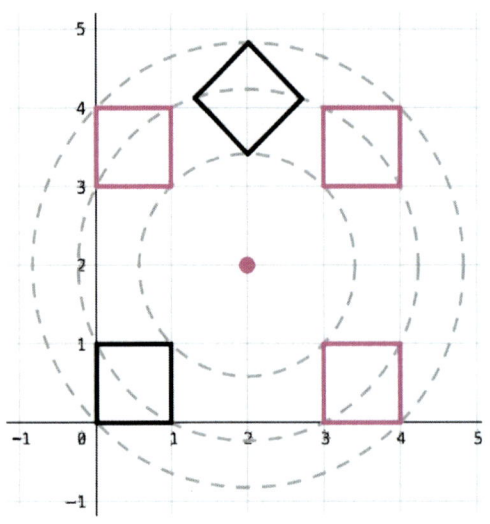

Task 6: The area of the square is 1 square unit. A square the same distance from the point (2,2) can be drawn anywhere on the circles (left), but the simplest are those drawn in pink:

Task 7: The line is parallel to the x-axis, with a midpoint at (6,2). Dividing it into thirds would segment the line at (4,2) and (8,2).

Task 8: The original points are at (1,1), (1,2) and (2,2). The reflected points are at (4,2), (5,1) and (5,2).

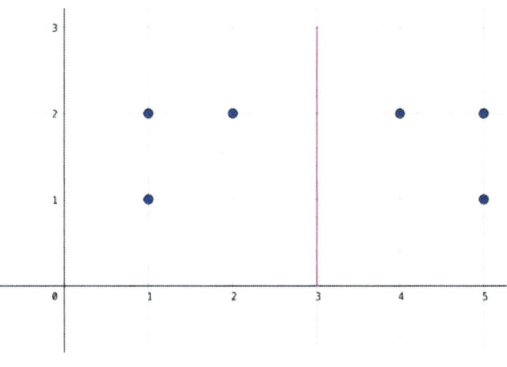

Task 9: The fourth triangle would create a white square in the centre. One way you could calculate the area of a triangle is by boxing it in as shown below and subtracting the area of the white triangles from the area of the rectangle. This would give you an area of 4 square units.

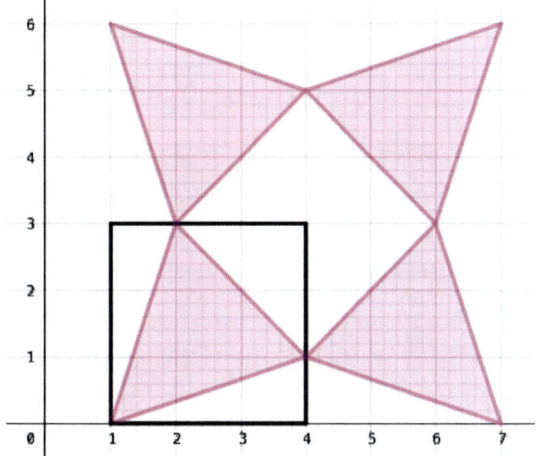

Task 10: A fourth triangle would include the additional point (6,4), and the eighth triangle in the sequence would have vertices at the points (8,4), (9,2), (10,4).

Nets

There are 11 ways to make a net for a cube:

There are eight ways to make a net for a square based pyramid:

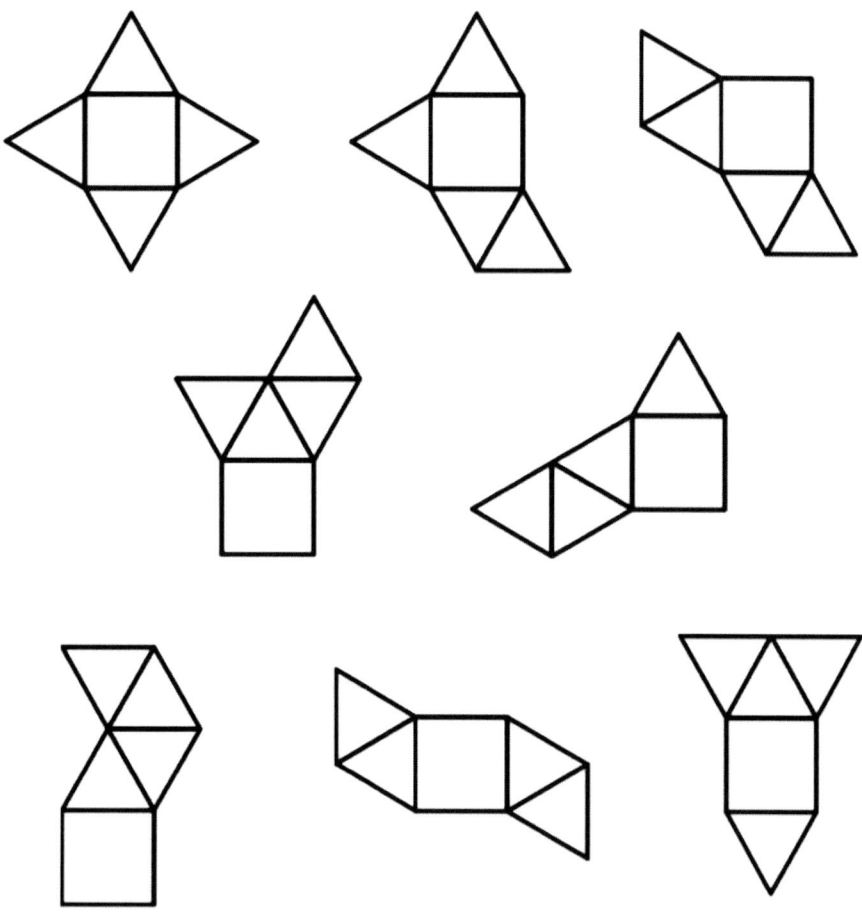

3-d Shape	Name	Faces	Edges	Vertices
	Pentagonal Prism	7	15	10
	Tetrahedron	4	6	4
	Pentagonal Pyramid	6	10	6
	Octahedron	8	12	6

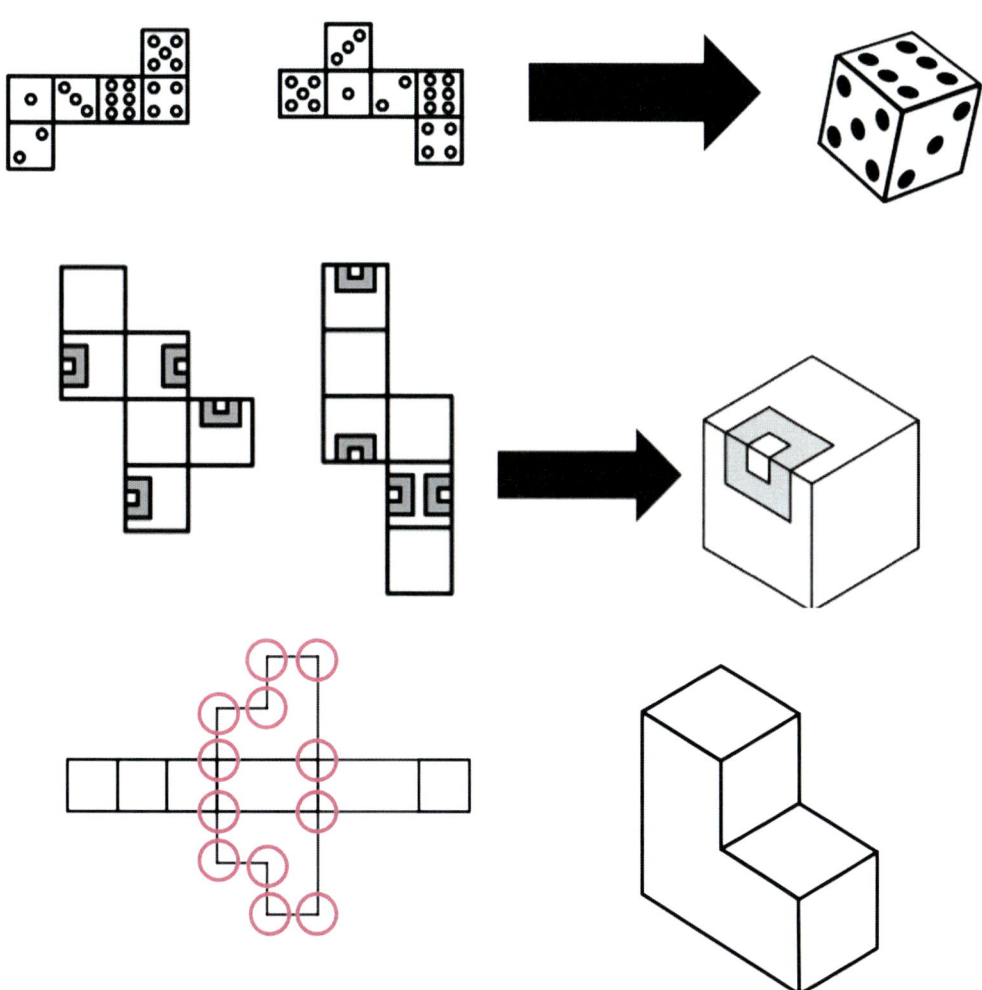

This shape has 12 vertices, 8 faces and 18 edges.

Angle Puzzles

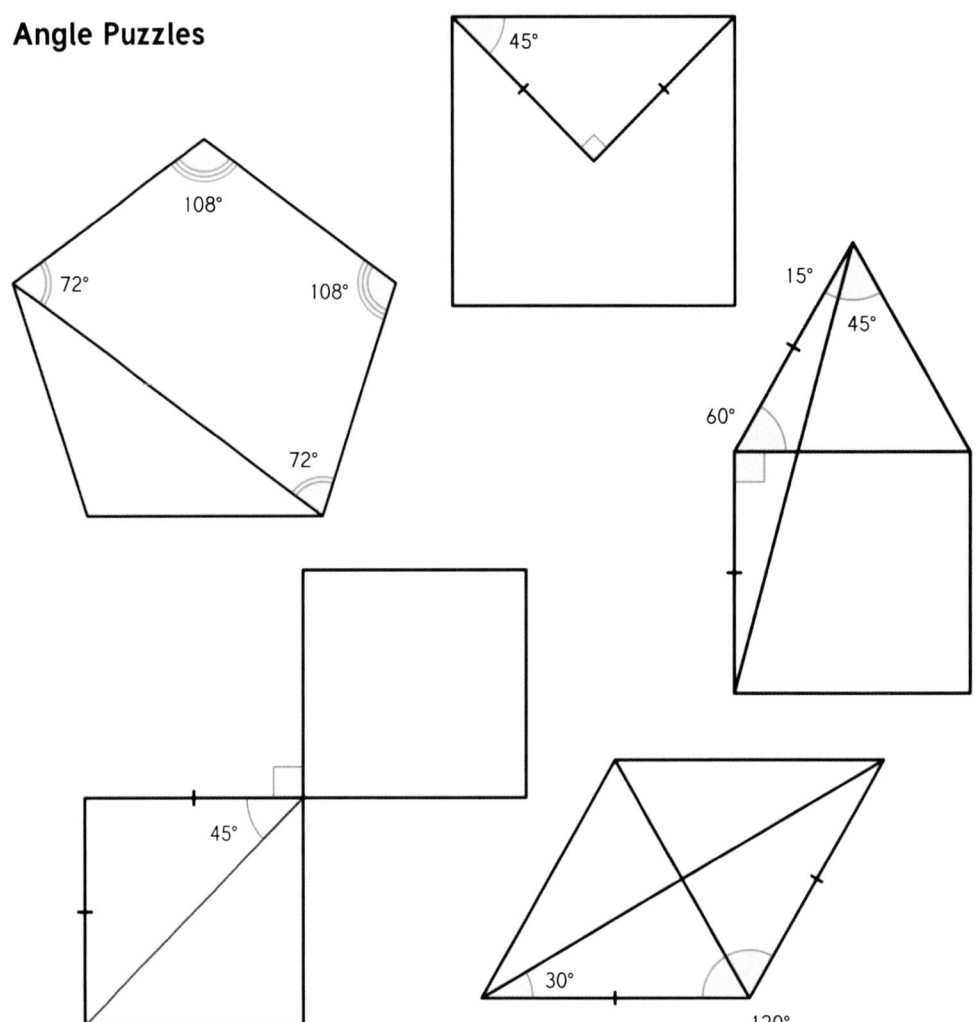

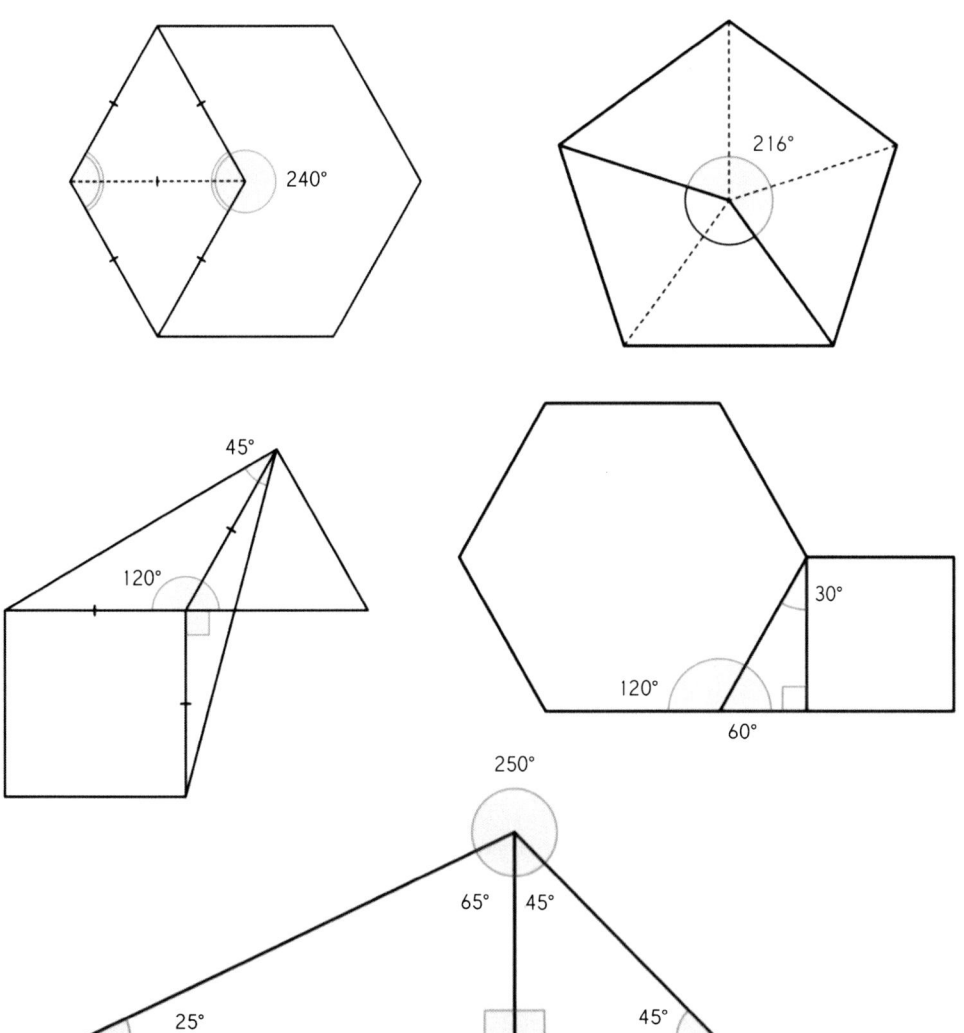

Area Puzzles

Task 1: The areas of each triangle is 3 square units, and the rectangle is 6 square units.

Task 2: Each rectangle has the same area of 4 square units.

Task 3: The shaded area for each shape is the same, but the amount of each shape that is shaded is different.

Task 4: Half of each shape is shaded (It's sometimes easier to focus on the white space rather than the shaded space).

Task 5: A half of each shape is shaded (if we assume that the triangle utilises a midpoint).

Task 6: The white triangles have areas of $8cm^2$ and $16cm^2$ respectively.

Task 7: Triangles with the same base and height have the same area, such as FBA and GDC, or GDB and FCA, or GCF and GBF.

Task 8: Both shapes have the same height, so the triangle must be twice the length of the square if they have the same area.

Task 9: The triangle that is part of both rectangles is half of each of them. Therefore they have the same area.

Task 10: The triangle has an area of $5cm^2$.